Research on the
PHYSICS
OF CANCER
A Global Perspective

Research on the
PHYSICS
OF CANCER

A Global Perspective

Editor

Bernard S. Gerstman, Ph.D.
Florida International University, USA

World Scientific

NEW JERSEY · LONDON · SINGAPORE · BEIJING · SHANGHAI · HONG KONG · TAIPEI · CHENNAI · TOKYO

Published by

World Scientific Publishing Co. Pte. Ltd.

5 Toh Tuck Link, Singapore 596224

USA office: 27 Warren Street, Suite 401-402, Hackensack, NJ 07601

UK office: 57 Shelton Street, Covent Garden, London WC2H 9HE

British Library Cataloguing-in-Publication Data

A catalogue record for this book is available from the British Library.

RESEARCH ON THE PHYSICS OF CANCER
A Global Perspective

ISBN 978-981-4730-25-9

In-house Editor: Christopher Teo

Typeset by Stallion Press
Email: enquiries@stallionpress.com

Printed in Singapore

Preface

Research on the Physics of Cancer: A Global Perspective

Bernard S. Gerstman

Chairman, Department of Physics
Florida International University
Miami, Florida, USA

Efforts have been made to cure and prevent cancer for centuries. These efforts received a considerable boost in 1971, when a "war" on cancer was declared. The outcomes of this war have been mixed. Treatments have been developed that have extended the lives of patients with some forms of cancer, but for other forms of cancer little progress has been made. Unfortunately, even the successful treatments usually have serious, horrible side-effects. Also discouraging, is that treatments rarely result in a complete cure in which all cancerous cells are killed or removed with no reoccurrence.

The steady but limited progress that has been made in treating cancer makes it imperative that we continue to refine medical treatments. However, the lack of ability to prevent or completely cure cancer is a clear indication that this is a complex disease. The use of the word "complex" is meant in both its general sense of being complicated, but also in the mathematical-physics sense of a system with multiple degrees of freedom and multiple interactions that interact in a non-linear fashion such that the whole is greater than the sum of the parts. For any complex system, external interventions often have unexpected and sometimes limited effects on the behavior of the system. The treatments for cancer display this condition. In addition, the ability of cancer cells to evolve and create resistance to external interventions creates enormous challenges to successful treatments. The complexity is further exacerbated by the fact that cancer displays multiscale behavior: molecular, cellular, and organismal.

In order to control the behavior of complex systems, two approaches can be employed. One approach is to apply external forces that are aimed at specific aspects of the behavior of the system. This approach has been used for decades to treat cancer, and many of the techniques have been introduced by physicists and are continuously refined. The other approach is to investigate the system to determine the details of the underlying, micro-interactions and create a mathematical model that describes how they interact to produce the macro-behavior. The macro-behavior of the system can then be controlled by targeting the micro-interactions. Fortunately, physicists have centuries of expertise in both approaches.

In recent years, physicists have turned their attention and expertise to understand and treat cancer. This volume reports on the work of several international teams that have devoted themselves to this task.

The chapter entitled, "Multiscale Modelling of Cancer Progression and Treatment Control: The Role of Intracellular Heterogeneities in Chemotherapy Treatment" by Mark A. J. Chaplain and Gibin G. Powathil, is a wonderful example of an investigation of cancer that combines a micro, mathematical approach with a macro treatment. Chaplain and Powathil use multiscale mathematical models on the sub-cellular, cellular, and microenvironmental levels to better understand the effects of chemotherapy on the growth and progression of cancer cells.

Claudia T. Mierke's chapter entitled, "Physical View on the Interactions Between Cancer Cells and the Endothelial Cell Lining During Cancer Cell Transmigration and Invasion" shines a new, physical light on cancer cell mechanics and the special role of the endothelium on cancer cell invasion. Mierke discusses the functional mechanism that cancer cells use to invade connective tissue and transmigrate through the endothelium to finally metastasize. Also discussed are approaches in which biophysical measurements can be combined with classical analysis approaches of tumor biology. These insights into physical interactions between cancer cells, the endothelium and the microenvironment may help to answer some "old" but still important questions in cancer disease progression.

The chapter by Zdenka Kuncic entitled, "Advances in Computational Radiation Biophysics for Cancer Therapy: Simulating Nano-Scale Damage by Low-Energy Electrons" reports on recent developments in computational radiation biophysics to improve cancer imaging and therapy. A significant advance in our ability to investigate the initial stages of radiation induced biological damage at the molecular level is achieved by modeling the interactions of very low energy electrons with water. This work suggests that radiation dose alone is insufficient to fully quantify biological damage and that radiation cross-fire may be an important clue to understanding the different observed responses of healthy cells and tumor cells to microbeam radiation therapy (MRT).

Alexander Iomin in his chapter, "Continuous Time Random Walk and Migration-Proliferation Dichotomy of Brain Cancer" uses the mathematical-physics framework of the random walk to analyze the underlying dynamics of glioma brain cancer. He uses fractional kinetics to explain the migration-proliferation dichotomy in the outer-invasive zone of glioma cancer cells. The aim is to determine how the migration-proliferation dichotomy influences the therapeutic effects of a radiofrequency electric tumor treatment field (TTF).

The chapter entitled, "Using Physics to Diagnose Cancer" by Veronica J. James discusses the use of X-ray diffraction as a tool for diagnosing cancer. X-ray diffraction was first used by Max von Laue and the team of William L. Bragg and William H. Bragg to investigate crystals of materials with periodic structures. This chapter explains how, over the decades, this physics technique has evolved into an important tool in medical diagnoses.

Roberto Chignola, Michela Sega, Sabrina Stella, Vladislav Vyshemirsky, and Edoardo Milotti in their chapter entitled, "From Single-Cell Dynamics to Scaling Laws in Oncology" report on their development of a mathematical, biophysical model of tumor biology. By using a multi-scale modeling approach, they show how a description of cancer at the cellular level leads to general laws obeyed by both in-vitro and in-vivo tumors. Importantly, each step of their model is validated by comparing simulation outputs with experimental data, which then allows the model to explore territories of cancer biology where current experimental techniques fail.

The chapter by Olivia Crociani, Andrea Becchetti, Duccio Fanelli, and Annarosa Arcangeli entitled, "Adhesion-Mediated Signalling in Cancer: Recent Advances and Mathematical Modelling" discusses a mathematical model designed to uncover the detailed connections between cellular apoptosis and hypoxia. This model has been constructed based upon experimental observations and shows how a dynamical switch can direct a cell between the normoxia and cellular death conditions. For certain levels of oxygen, cells can cross the transition line and head towards the normoxia regime by implementing point mutations that affect the p53 production and activation rate, with the involvement of K+ ion homeostasis. This work is relevant to the recently identified novel signalling pathway centered on hERG1 channels and integrins in colorectal cancer (CRC) that involves the p53 protein and regulated by hERG1 K+ channels. The central role played by hERG1 in CRC angiogenesis suggests that targeting hERG1 may be an effective therapeutic option in patients with advanced CRC.

Kathrin Zeller and Staffan Johansson in their chapter entitled, "Common and Diverging Integrin Signals Downstream of Adhesion and Mechanical Stimuli and Their Interplay with Reactive Oxygen Species" also report on work involving the integrin family of adhesion receptors and how the signals they induce are altered in tumor cells. They discuss how different integrin-dependent signals are generated during cell adhesion and by physical forces acting on cells. They describe how reactive oxygen species are integral parts of integrin signaling and highlight important questions in the field that may improve our understanding of integrins and their role in the development of cancer.

The chapter entitled, "Can Mathematical Models Predict the Outcomes of Prostate Cancer Patients Undergoing Intermittent Androgen Deprivation Therapy?" by R. A. Everett, A. M. Packer, and Y. Kuang report on their work on a mathematical model which used measurements of patient testosterone levels to accurately fit measured serum prostate specific antigen (PSA) levels. Like the normal prostate, most tumors depend on androgens for proliferation and survival but often develop treatment resistance. Hormonal treatment causes many undesirable side effects which significantly decrease the quality of life for patients. They test the model's predictive accuracy using only a subset of the data to find parameter values. The results are compared with those of an existing piecewise linear model which does not use testosterone as an input. They raise the possibility that a simpler model may be more beneficial for predictive use compared to a more biologically insightful model.

James P. Brody, in his chapter entitled, "The Age Specific Incidence Anomaly Suggests that Cancers Originate During Development" discusses the anomaly that the incidence for many forms of cancer rises with age to a maximum and then decreases, rather than monotonically increasing with age. By focusing on the age related incidence of colon cancer, he explains how this leads to the interpretation that two subpopulations exist in the general population: a susceptible population and an immune population, and that colon tumors will only occur in the susceptible population. This analysis is consistent with the developmental origins of disease hypothesis and generalizable to other common forms of cancer.

The chapter entitled, "Cancer — Pathological Breakdown of Coherent Energy States" by Jiri Pokorny, Jan Pokorn, Jitka Kobilkov, Anna Jandov, Jan Vrba, and Jan Vrba, Jr. discusses how mitochondria in eukaryotic cells produce energy and form conditions for excitation of oscillations in microtubules and how mitochondrial dysfunction (the Warburg effect) in cancer development excludes the afflicted cell from the ordered multicellular tissue system. They suggest that cancer treatment should target mitochondrial dysfunction, but note that asbestos fibers present a special problem for treatment.

Yu Jiang and Weiqun Wang discuss how the generally good advice that a person should keep their body weight down is also applicable to cancer in their chapter entitled, "Potential Mechanisms of Cancer Prevention by Weight Control". Weight control via dietary caloric restriction and/or physical activity as a means of cancer prevention has been demonstrated in animal models. However, the underlying mechanisms are not fully understood. The prevalence of obese or overweight individuals makes the understanding of the underlying mechanisms of the connections between weight control, endocrine change, and cancer risk critically important. They discuss multiple factors and multiple signaling cascades that may be important, especially for Ras-MAPK-proliferation and PI3K-Akt-anti-apoptosis.

CONTENTS

Chapter 1

Multiscale Modelling of Cancer Progression and Treatment Control: The Role of Intracellular Heterogeneities in Chemotherapy Treatment

Mark A. J. Chaplain* and Gibin G. Powathil[†,‡]

Division of Mathematics, University of Dundee
Dundee, DD1 4HN, Scotland
*chaplain@maths.dundee.ac.uk
†gibin@maths.dundee.ac.uk;
†g.g.powathil@swansea.ac.uk

Cancer is a complex, multiscale process involving interactions at intracellular, intercellular and tissue scales that are in turn susceptible to microenvironmental changes. Each individual cancer cell within a cancer cell mass is unique, with its own internal cellular pathways and biochemical interactions. These interactions contribute to the functional changes at the cellular and tissue scale, creating a heterogenous cancer cell population. Anticancer drugs are effective in controlling cancer growth by inflicting damage to various target molecules and thereby triggering multiple cellular and intracellular pathways, leading to cell death or cell-cycle arrest. One of the major impediments in the chemotherapy treatment of cancer is drug resistance driven by multiple mechanisms, including multi-drug and cell-cycle mediated resistance to chemotherapy drugs. In this article, we discuss two hybrid multiscale modelling approaches, incorporating multiple interactions involved in the sub-cellular, cellular and microenvironmental levels to study the effects of cell-cycle, phase-specific chemotherapy on the growth and progression of cancer cells.

1. Introduction

Even with many important clinical and technological advancements in detecting and treating cancer, cure and control of many forms of cancer remain the greatest challenge to clinicians and scientists. In most cases, chemotherapy is used alone or in combination with other anticancer treatments such as radiotherapy and surgery to control a growing tumour. However, drug resistance driven by multiple mechanisms, including multi-drug and cell-cycle mediated resistances to chemotherapy drugs continues to be a major barrier for the treatment failure in human malignancies.[1,2] Several recent experimental studies have indicated the fundamental role of intratumoural heterogeneity as a driving source for the resistance to multiple chemotherapeutic drugs.[3,4] One of the major reasons for this intratumoural heterogeneity is the intracellular perturbations in biochemical kinetics and heterogeneity

---

‡*Current address*: Department of Mathematics, Swansea University, Swansea, SA2 8PP, UK.

in the tumour microenvironment that seriously impair the drug efficacy.[1] Hence, understanding various mechanisms involved in the development of drug resistance and, devising drugs and protocols to target these mechanisms are significant steps in overcoming drug resistance, where clinically driven computational models can play an important role.[5,6]

The two main factors that contribute to the intra-tumoural heterogeneity are internal cell-cycle dynamics and the surrounding oxygen concentration. The cell-cycle mechanism through which the cells duplicate consists of several transition phases of varying lengths and check points and is mainly divided into four phases. As most of the chemotherapeutic drugs that are administered to treat human malignancies are cell-cycle phase specific, they spare some of the cells that are in the non-targeted phase, causing a cell-cycle mediated drug resistance.[2] Cell-cycle dynamics are also further influenced by the external microenvironmental conditions, especially the availability of oxygen. Experimental evidence shows that hypoxia (lack of oxygen) can upregulate the expressions of some of the cyclin dependent kinase inhibitors such as p21 and p27, resulting in a prolonged cell-cycle time or even cell-cycle arrest.[7,8] This further contributes to the cell-cycle heterogeneity and cell-cycle phase specific drug resistance. Here, we discuss a multiscale mathematical model, incorporating some of these cellular heterogeneities to understand and study their role in inducing chemotherapeutic drug resistance.

The multiscale complexity of cancer progression warrants a multiscale modelling approach to produce truly clinically useful and predictive mathematical models. Previously, Powathil *et al.*[5] developed a multiscale mathematical model of chemotherapy treatment, incorporating cell-cycle mediated intracellular heterogeneity and external oxygen heterogeneity to study the effects of cell-cycle, phase-specific chemotherapy and its combination with radiation therapy.[9] It has been shown that an appropriate combination of cell-cycle specific chemotherapeutic drugs with radiation delivery could effectively be used to control tumour progression. There have been several mathematical and computational modelling approaches developed to study the occurrence of drug resistance.[10,11] These approaches help to understand and to some extent analytically quantify various biological processes. Furthermore, it can also be used as a tool to analyse and design drug development experiments and clinical trials. In this article, we discuss the multiscale mathematical model developed by Powathil *et al.*[5] and two different computational approaches to implement the developed model. Further, we use it to study the effects of cell-cycle phase-specific chemotherapeutic drugs on a growing tumour population with intratumoral heterogeneities.[6]

2. Modelling Cancer Growth: Multiple Scales Involved

Cancer growth is a complicated multiscale disease involving many interrelated processes that occur across a wide range of spatial and temporal scales, from the intracellular level to the tissue level. Consequently, a multiscale modelling approach is needed to capture the key processes that are occurring at these different spatial and temporal scales and couple them in an appropriate manner. Here we discuss a hybrid multiscale model developed by Powathil *et al.*[5] that analyses the spatio-temporal dynamics at the level of individual cells, linking individual cell behavior with the macroscopic behavior of cell/tissue organisation and the microenvironment. The model captures the intracellular molecular dynamics of the cell-cycle pathway and the changes in oxygen dynamics within the tumour microenvironment. It is then used to study the impact of oxygen heterogeneity on the spatio-temporal patterning of the cell distribution and their cell-cycle status.[5,9]

The growth and progression of a solid tumour mass depends critically on the responses of the individual cells that constitute the entire tumour mass. The evolution of each individual cancer cell and its decisions to grow, divide, remain inactive or die are usually influenced by the local micro-environmental conditions at the location occupied by any particular cell within the tumour and intracellular interactions, including the intracellular cell-cycle dynamics. Moreover, these cellular responses are actively influenced by various extracellular signals from neighbouring cells as well as its dynamically changing microenvironment. As discussed in Powathil *et al.*,[5] the growth and proliferation of each cancer cell is determined by its own internal cell-cycle mechanism and is incorporated using a set of ordinary differential equations. This internal cell-cycle dynamics are further influenced by the changing surrounding oxygen concentration which is modelled through the activation of HIF pathway (hypoxia inducible factor pathway) linking the microenvironment to intracellular cell-cycle pathway.

The HIF pathway in usually implicated in several hypoxia related events within a growing tumour such as the production of metastatic phenotypes with increased mutation rates, increased secretion of angiogenic factors, less apoptosis and an up-regulation of various pathways involved in the metastatic cascade.[12] The hypoxia inducible transcription factor-1 is composed of two subunits, HIF-1 α and HIF-1 β, both of which are required for its transcription activation function. Under normoxic conditions, the rapidly produced HIF-1α is degraded immediately by the actions of proline hydroxylase and pVHL. However, under hypoxic conditions HIF-1 α escapes degradation and its level increases rapidly. This further activates the expression of various genes, triggering various intra- and intercellular pathways including the expressions of cyclin dependent kinase inhibitors p21 and p27 pathways, affecting the cell-cycle dynamics.[7,13] The dynamical changes in the tumour microenvironment due to the variations in oxygen concentration are modelled using partial differential equations. The developed model can be then used to analyse cellular heterogeneities due to various internal and external factors and its role in a cell's response to chemotherapy treatment.

2.1. *Intracellular heterogeneities*: *Modelling the cell-cycle dynamics*

Most of the complex cellular processes that are involved in cancer progression such as proliferation, cell division and DNA replication are regulated by the cell-cycle. The cell-cycle is controlled by a complex hierarchy of metabolic and genetic networks with several transition phases of varying lengths and check points.[14] The cell-cycle can be divided into four main phases: S-phase where DNA synthesis occurs, G2-phase during which proteins required for mitosis are produced, M-phase where mitosis and separation occur and G1-phase where proteins necessary for S-phase progression are accumulated.[15] Additionally, cells may sometimes exit from the cell-cycle and enter a phase of quiescence or relative inactivity called the G0-phase or resting phase.[14] The cell-cycle dynamics within a mammalian cell are regulated mainly by a family of cyclin dependent kinases (Cdk), whose activity is primarily dependent on association with a regulatory protein called cyclin.[16] Additionally, the progression of cell-cycle dynamics is affected by several intracellular and extracellular factors such as Cdk inhibitors that can act as negative regulators of the cell-cycle and tumour microenvironment.[15] A few specific examples of Cdk inhibitors include the proteins p16, p15, p21 and p27. Some of the extrinsic factors that can influence the cell-cycle mechanism include nutrient supply, cell size, temperature and cellular oxygen concentration.[14]

Here we use a cell-based modelling approach to study the growth and progression of a cancer cell mass. The evolution of each cancer cell is based on the decisions made by the cell-cycle mechanism within the cell and we further assume that this contributes to the intracellular heterogeneities. To model the cell-cycle dynamics within each cell, we use an adapted version of a very basic model[5] originally developed by Tyson and Novak[17,18] that includes only the interactions which are considered to be essential for cell-cycle regulation and control. The models by Tyson and Novak[17,18] describe the cell-cycle as a hysteresis loop with two self-maintaining stages while the transitions between these two stages are determined by the changing cell mass during the division. They used kinetic relations between various chemical processes to study the transitions between two main steady states, G1 and S-G2-M of the cell-cycle, which is (in their model) controlled by changes in cell mass. Although, Tyson and Novak have subsequently introduced a much more sophisticated model for the mammalian cell-cycle,[19] for simplicity we have opted to use the six variable model to simulate the cell-cycle. Moreover in the adapted model, we have used the equivalent mammalian proteins stated in Tyson and Novak's paper, namely the Cdk-cyclin B complex [CycB], the APC-Cdh1 complex [Cdh1], the active form of the p55cdc-APC complex [p55cdc$_A$], the total p55cdc-APC complex [p55cdc$_T$], the active form of Plk1 protein [Plk1] and the mass of the cell [mass].[5,9] Using the kinetic relations, the evolution of the concentrations of these variables are modelled using the following system of six ODEs (further details concerning the kinetic interactions can be found in Tyson and Novak's papers.[17,18])

$$\frac{d[\text{CycB}]}{dt} = k_1 - (k_2' + k_2''[\text{Cdh1}] + [\text{p27/p21}][\text{HIF}])[\text{CycB}], \tag{1}$$

$$\frac{d[\text{Cdh1}]}{dt} = \frac{(k_3' + k_3''[\text{p55cdc}_\text{A}])(1 - [\text{Cdh1}])}{J_3 + 1 - [\text{Cdh1}]} - \frac{k_4[\text{mass}][\text{CycB}][\text{Cdh1}]}{J_4 + [\text{Cdh1}]}, \tag{2}$$

$$\frac{d[\text{p55cdc}_\text{T}]}{dt} = k_5' + k_5'' \frac{([\text{CycB}][\text{mass}])^n}{J_5^n + ([\text{CycB}][\text{mass}])^n} - k_6[\text{p55cdc}_\text{T}], \tag{3}$$

$$\frac{d[\text{p55cdc}_\text{A}]}{dt} = \frac{k_7[\text{Plk1}]([\text{p55cdc}_\text{T}] - [\text{p55cdc}_\text{A}])}{J_7 + [\text{p55cdc}_\text{T}] - [\text{p55cdc}_\text{A}]}$$
$$- \frac{k_8[\text{Mad}][\text{p55cdc}_\text{A}]}{J_8 + [\text{p55cdc}_\text{A}]} - k_6[\text{p55cdc}_\text{A}], \tag{4}$$

$$\frac{d[\text{Plk1}]}{dt} = k_9[\text{mass}][\text{CycB}](1\text{-}[\text{Plk1}]) - k_{10}[\text{Plk1}], \tag{5}$$

$$\frac{d[\text{mass}]}{dt} = \mu[\text{mass}]\left(1 - \frac{[\text{mass}]}{m_*}\right), \tag{6}$$

where k_i are the rate constants and the values are chosen in proportion to those in Tyson and Novak so that the time scale is relevant to a mammalian cell-cycle.[5] Other parameters used in the system are J_i, [Mad] and [p27/p21].[5] The effects of changes in oxygen dynamics are included into the system through the activation and inactivation of HIF pathway which further results in changes in cell-cycle length. Here, we have assumed that HIF-1 α concentration at a cellular position, which is normally inactive ([HIF] = 0), is activated ([HIF] = 1) if the oxygen concentration at that position falls below 10%. The cell-cycle inhibitory effect of p21 or p27 genes expressed through the activation of HIF-1 α is incorporated into the equation governing our generic Cyclin-CDK dynamics, using an additional decay term proportional to the concentration of p27/p21 (which is considered here as constant).[5,20] A cell is assumed to divide when the concentration of Cdk-cyclin B complex [CycB] crosses a specific threshold value [CycB]$_{th}$ which is assumed to be 0.1, from above, and then the mass, [mass] is halved. To introduce a random growth rate for individual cells which in turn introduces cell-cycle heterogeneity in the population, we consider a varying growth rate μ. The rest of the parameter values of the cell-cycle model can be found in Powathil *et al.*[5]

Figure 1 shows the changes in various protein concentrations that have been included in the current cell-cycle model for one single automaton cell. Every cell in this multiscale model has a similar cell-cycle dynamics built-in which further control the division and cell-cycle phases of the respective cells. In this representative figure (adapted from Powathil *et al.*[5]), a cell undergoes division constantly as long as there is enough space to divide and the surrounding microenvironment is favourable for its division. However, as soon as all its neighbouring spaces are occupied, the cell moves to a resting phase where the concentrations are maintained at a constant level.

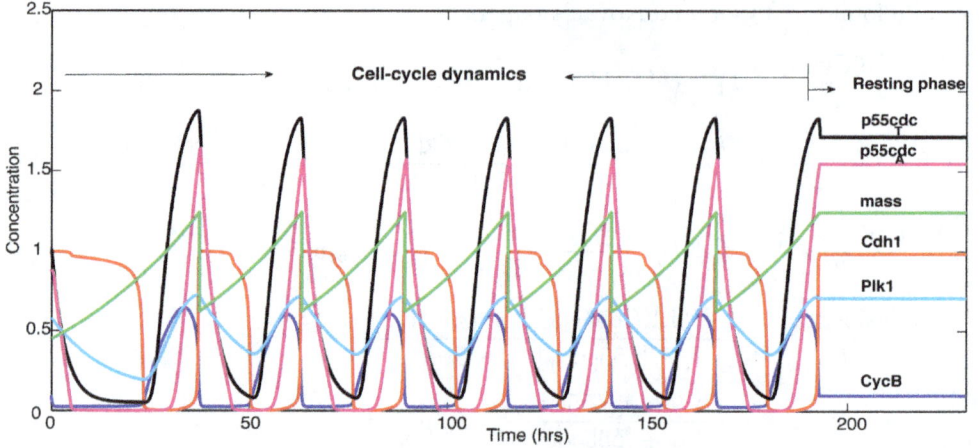

Fig. 1.   Plot of the concentration profiles of the various intracellular proteins and the cell-mass over a period of 200 hours for one automaton cell in the model. This is obtained by solving the system of equations, (1)–(6), with the relevant parameter values. Adapted from Powathil et al.[5]

2.2. Microenvironment heterogeneities: Modelling the oxygen dynamics

The growth of individual tumour cells as well as the entire tumour mass is externally influenced by its surrounding microenvironment. In particular the local availability of nutrients such as oxygen. The effects of a dynamically changing microenvironment introduced by incorporating oxygen dynamics, is modelled using a partial differential equation.[5,9] Here, oxygen is assumed to be supplied from a random distribution of blood vessels (vascular cross sections in 2D) with a density of $\phi_d = N_v/N^2$, where N_v is the number of vessel cross sections in the 2-dimensional domain (of area N^2).[5] This is a reasonable assumption if the blood vessels are assumed to be perpendicular to the tissue cross section of interest and there are no branching points through the plane of interest.[21,22] The temporal dynamics of these vessels are ignored at present, assuming the growth of tumour cells is much faster than that of the vessels within the time frame of interest. Denoting by $K(x,t)$ the oxygen concentration at position x at time t, then its rate of change can be expressed as,

$$\frac{\partial K(x,t)}{\partial t} = \nabla \cdot (D_K(x)\nabla K(x,t)) + r(x)m(x) - \phi K(x,t)\text{cell}(x,t) \qquad (7)$$

where $D_K(x)$ is the diffusion coefficient and ϕ is the rate of oxygen consumption by a cell at position x at time t (cell$(x,t) = 1$ if position x is occupied by a cancer cell at time t and zero otherwise). Here, $m(x)$ denotes the vessel cross section at position x ($m(x) = 1$ for the presence of blood vessel at position x, and zero otherwise) and $r(x)$ describes rate of oxygen supply.[5] This equation is solved using no-flux boundary conditions and an initial condition.[23] Figure 2 shows a representative profile of the spatial distribution of oxygen concentration after solving equation (7)

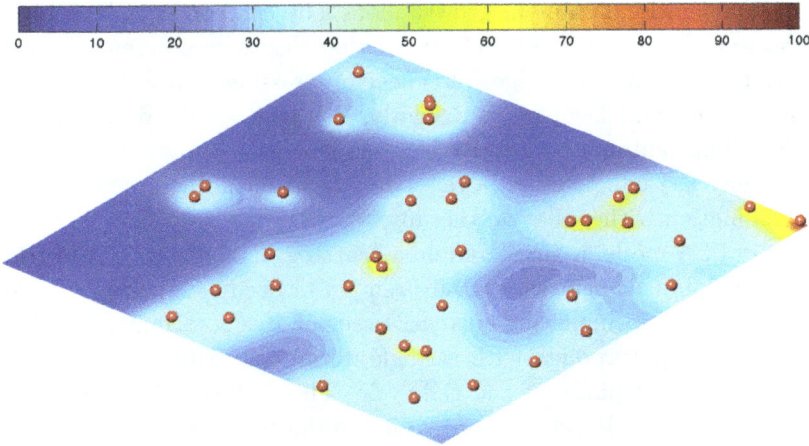

Fig. 2. Plot showing the concentration profile of oxygen supplied from the vasculature in the local tissue. The red coloured spheres represent the blood-vessel cross-sections and the colour map shows the percentages of oxygen concentration. Adapted from Powathil *et al.*[9]

with relevant parameters.[5] Furthermore, It is observed that lack of an adequate supply of oxygen (hypoxia) can upregulate some of the cell-cycle inhibitory proteins such as p27 and p21 which could interfere with the cell-cycle, eventually taking the cell either to a resting phase or inducing a cell-cycle arrest.[24,7] These effects are introduced into the cell-cycle dynamics through the equation governing the changes of Cdk-cyclin B complex (cf. Eq. (1)).

3. Implementation of the Multiscale Model

The tissue-scale dynamics of the oxygen concentration outlined above can be linked to the sub-cellular and cellular changes through two different modelling approaches, namely, (i) a hybrid multiscale cellular automaton framework[5,9] and (ii) a multiscale cellular Potts modelling approach using the CompuCell3D framework.[6] In both these modelling approaches, the computational domain contains three different components that are required to simulate the multiscale model. These are: (1) the cancer cells whose spatio-temporal evolution is controlled by internal cell-cycle dynamics and the external microenvironment; (2) the oxygen concentration distribution and (3) cross-sections of blood vessels from where the oxygen and chemotherapeutic drugs are supplied to the domain of interest. A detailed explanation of these modelling approaches are outlined below.

3.1. *The hybrid multiscale cellular automaton framework*

Cellular automaton (CA) modelling has been used very extensively to model various aspects of tumour development and progression.[25,21,26,27] Some examples for such studies include multiscale tumour growth models by Alarcon *et al.*,[28] Ribba *et al.*[29] and Gerlee and Anderson.[30] A brief review that discusses different CA modelling approaches to study various stages of cancer progression can be found in a review article by Moreira and Deutsch.[31] Recently, Powathil *et al.*[5] developed a hybrid multiscale cellular automaton approach to model cancer progression and used the model to study the effects of cell-cycle dependent chemotherapeutic drugs alone and in combination with radiation therapy.[5,9]

The hybrid CA model is simulated on a spatial grid of size 100×100 grid points and each automaton element whether it is empty or occupied, has a physical size of $l \times l$, where $l = 20\,\mu$m, simulating a cancer tissue of area $2 \times 2\,$mm^2. The CA begins as a new grid of empty points with a single initial cell (a blue cell) at the centre of the grid in the G1-phase of the cell-cycle. This initial cell divides repeatedly following its internal cell-cycle dynamics and produces a cluster of cells on a square lattice (no-flux boundary conditions are imposed). The entire multiscale model is simulated over a certain period of time and a vector containing all cell positions and intracellular protein levels for each cell are updated accordingly. The oxygen dynamics are simulated using a finite difference scheme at every simulation time step and the corresponding oxygen concentration levels are updated. The cell-cycle phases are determined using the concentration levels of [CycB]. If [CycB] is greater than a specific threshold (i.e. 0.1) the cell is considered to be in the S-G2-M phase (green cell) and if it is lower than this value, the cell is in the G1-phase. If the cyclinB-cdk complex concentration [CycB] crosses this threshold from above, the cell undergoes cell division, its mass [mass] is halved. Alternatively, a cell may enter into a resting phase if the dividing cell's neighbourhood has no space for the new daughter cell. Alternatively, if division takes place, the new cell is placed into the G1-phase of the cell-cycle and is assigned a value for its proliferation rate μ randomly from the range of values of μ. If there is more than one empty space with the same oxygen concentration level, a position is chosen randomly. The position of the new daughter cells is determined by alternating Moore and Von Neumann neighbourhoods to avoid generating cell distribution patterns matching the specific neighbourhood[5,9] (this creates symmetric "circular" masses of cancer cells as opposed to square or diamond shapes). As the cancer cells proliferate, the oxygen demand increases making an imbalance between supply and demand which will eventually create a state where the cells are deprived of oxygen. If the oxygen concentration falls below 10% the cells are assumed to be hypoxic and the hypoxic cells that are in G1-phase are represented by rose colour coded cells while hypoxic S-G2-M cells and hypoxic resting cells are denoted by the colours yellow and silver, respectively. Figure 3 shows the distribution of cells in various cell-cycle phases at three different time points. The simulation time step for the both the CA model and the oxygen dynamics is taken as T $= 0.001$ hr as it gives a oxygen diffusion constant of $2 \times 10^{-5}\,$cm^2/s with appropriate diffusion length scale L of $100\,\mu$m. Further details of the model can be found in Powathil *et al.*[5]

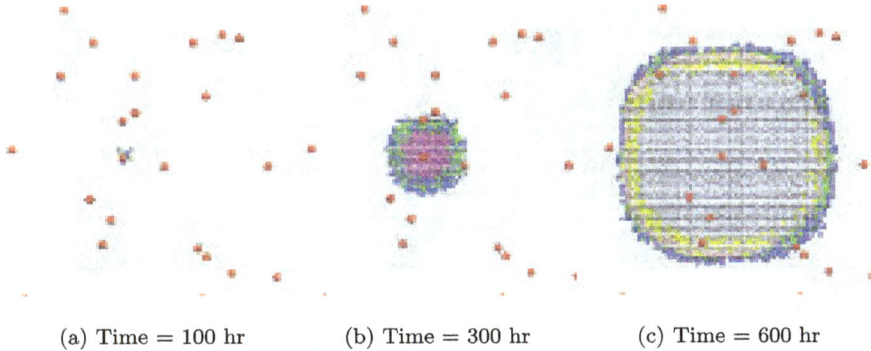

(a) Time = 100 hr   (b) Time = 300 hr   (c) Time = 600 hr

Fig. 3.   (Color online) Plots of the spatial distribution of the cells in different stages of the cell-cycle which are G1 (blue), S-G2-M (green), resting (magenta), hypoxic cells in G1 (rose), hypoxic cells in S-G2-M (yellow) and hypoxic cells in resting (silver) at times (a) 100 hr, (b) 300 hr and (c) 600 hr.

3.2.   *The multiscale cellular potts model*: *CompuCell3D framework*

An alternative approach to modelling such complex multiscale problems is by using a multiscale cellular Potts model or the Glazier-Graner-Hogeweg (GGH) approach. The GGH model contains description of objects, such as cells and fields, inter-actions with the cellular properties that evolve with respect to time and space and are modelled with the help of various initial conditions.[32] Each cell is a col-lection of lattice pixels having the same index marker and is represented as spa-tially extended domains on a fixed lattice. We used the CompuCell3D framework developed by Glazier *et al.* (see http://www.compucell3d.org for details) to simu-late the previously described multiscale model.[6] The multiscale model is simulated using a 2-dimensional lattice of size 300×300 pixels in the x- and y-directions with an initial configuration of single cells surrounded by a number of blood ves-sel cross sections. Similar to the previous approach, the division of tumour cells is driven by the cell-cycle dynamics modelled using the kinetic equations (1)–(5). The set of ODEs governing the cell-cycle dynamics is incorporated into the Compucell3D framework in each Monte Carlo time step (mcs) using Bionetsolver. Bionetsolver is a C++ library that permits easy definition of sophisticated models coupling reaction-kinetic equations described in SBML with the defined cells for execution in CompuCell3D.[32] Bionetsolver makes use of the SBML ODE Solver Library to implement reaction-kinetic network dynamics which can regulate the cell-cycle dynamics for each tumour cell within the domain. Cellular growth is incor-porated into the model by incrementing the cell target volume in every mcs during growth phases at a constant rate of 0.5 times the current cell volume. Division is assumed to occur when the concentration of [CycB] crosses the threshold value from above. However, since here we are using a growing volume, the cell-cycle dynamics are simulated using the equations (1)–(5), using [volume] instead of [mass]. The parameter values of the cell-cycle model are scaled in such a way that each mcs

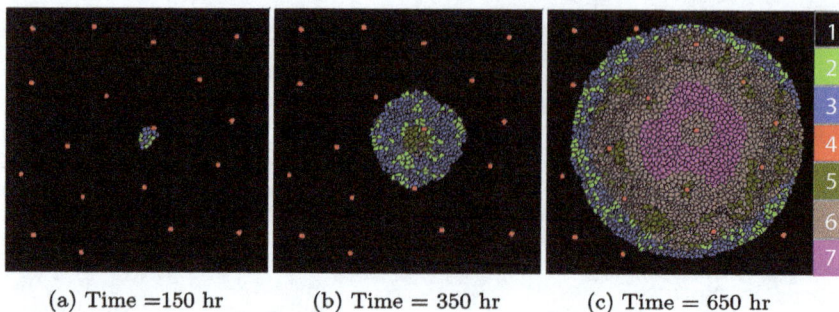

(a) Time =150 hr   (b) Time = 350 hr   (c) Time = 650 hr

Fig. 4.   Plots of the spatial evolution of tumour cells in different phases of the cell-cycle at times (a) 150 hr, (b) 350 hr and (c) 650 hr. The colour legend shows the types of the tumour cells; 1 - medium, 2 - G2 phase, 3 - G1 phase, 4 - vessel cross sections, 5 - hypoxic G2 phase, 6 - hypoxic G1 phase and 7 - resting cells. Adapted from Powathil *et al.*[6]

step corresponds to 1 hour and hence a cell has an average cell-cycle length of 25–35 hours. The evolution of oxygen concentration is incorporated into the Com-puCell3D as a diffusive chemical field that follows the respective PDE described previously. The parameters are taken from Powathil *et al.*[5] and for consistency, the diffusion equation is simulated 1000 times in every 1 mcs to achieve a similar time-scale of 0.001 hr.[6] Figure 4 shows the spatial evolution of tumour cells. The colour of the tumour cells indicate their cell-cycle phase position and the microenvironment status.

4. Modelling the Effects of Chemotherapy

Chemotherapy is one of the most common therapeutic options for cancer treat-ment, either alone or in combination with other therapies (multimodality). Chemotherapeutic drugs act on rapidly proliferating cells targeting the different cell-cycle phases and check points. In cancer, Cdks, the proteins responsible for the activation of the cell-cycle, are over-expressed while cell-cycle inhibitory proteins are under-expressed which results in a malfunctioning in the regulation of the cell-cycle, and eventually leads to a promotion of uncontrolled growth. The rationale behind cell-cycle, phase-specific chemotherapy is to target those proteins that are over-expressed during various stages of cancer progression, inducing an inhibitory effect and thus controlling cell growth. One of the major issues that affects the delivery and effectiveness of chemotherapeutic drugs is the occurrence of cell-cycle mediated drug resistance.[2] This may be due to the presence of functionally heterogeneous cells and cell subpopulations, and can be addressed to some extent by using combinations of chemotherapy drugs that target different phases of the cell-cycle kinetics.[2]

We are interested in studying the effects of cell-cycle based chemotherapeutic drugs on cancer cells and cancer cell subpopulations with varying drug sensitivi-ties. We model the spatio-temporal evolution of cell-cycle specific chemotherapeu-tic drugs using a similar partial differential equation as that governing the oxygen dynamics. Hence, denoting by $C_i(x, t)$ the concentration of chemotherapeutic drug type i, its spatio-temporal evolution is given by the equation:

$$\frac{\partial C_i(x, t)}{\partial t} = \nabla \cdot (D_{ci}(x)\nabla C_i(x, t)) + r_{ci}(x)m(x) - \phi_{ci}C_i(x, t)\text{cell}(x, t) - \eta_{ci}C_i(x, t)$$

(8)

where $D_{ci}(x)$ is the diffusion coefficient of the drug, ϕ_{ci} is the uptake rate by a cell (assumed to be zero), r_{ci} is the drug supply rate by the pre-existing vascular network and η_{ci} is the drug decay rate.[5,6] As similar to that of equation governing the oxygen concentration, this PDE is incorporated into the CompuCell3D as diffusive chemical field and simulated using the parameters values found in Powathil *et al.*[5] To study the effects of multiple phase-specific chemotherapy, we consider two types of phase-specific chemotherapeutic drugs that are either G1 specific or S-G2-M specific, delivered at a same rate. Furthermore, chemotherapeutic drugs are assumed to be effective in killing a cell, if its average concentration at the location of that specific cell is above a fixed threshold value and below which the drug has no effect on any cells. In the following subsections, we study the effects of cell-cycle based chemotherapeutic drugs on a growing tumour using the CompuCell3D framework hybrid multiscale computational model.

4.1. *Homogenous population model: The effects of chemotherapy*

In this section, we study we effects of cell-cycle based chemotherapy on a population of homogeneously growing tumour cells with similar cell-cycle dynamics (i.e. the same cell-cycle time under favourable conditions) but with intracellular and microenvironmental heterogeneities. Figure 4 shows the spatio-temporal evolution of a solid tumour mass with a homogenous cell population in the absence of chemotherapy. As illustrated in the figure, the colours of tumour cells indicate the cell-cycle position and oxygenation status of each individual cell. The spatial distribution of the tumour cells shows the development of the proliferating rim around the boundary of the growing tumour as the internal cells become hypoxic due to the increased consumption of oxygen supplied from the blood vessels.

To study the effects of cell-cycle, phase-specific chemotherapeutic drugs on the growing tumour, two doses of cell-cycle phase-specific drugs that act on cells that are either in G1-phase or S-G2-M phase are delivered at a same rate at times 500 hours and 550 hours. A representative figure showing the spatio-temporal evolution of cancer cells when the tumour mass is treated with two doses of G1 drugs and G2 drugs is given in Fig. 5. Figure 6 shows and compares the total number of tumour cells during the therapy. As previously shown by Powathil *et al.*,[5] the results indicate that the choice and sequencing of different types of chemotherapeutic drugs can significantly affect the spatial distribution and the cytotoxic cell-kill of cancer cells. Furthermore, it has been shown that various factors such as the spatial distribution of cancer cells, the correct sequencing of chemotherapeutic drugs, and intracellular

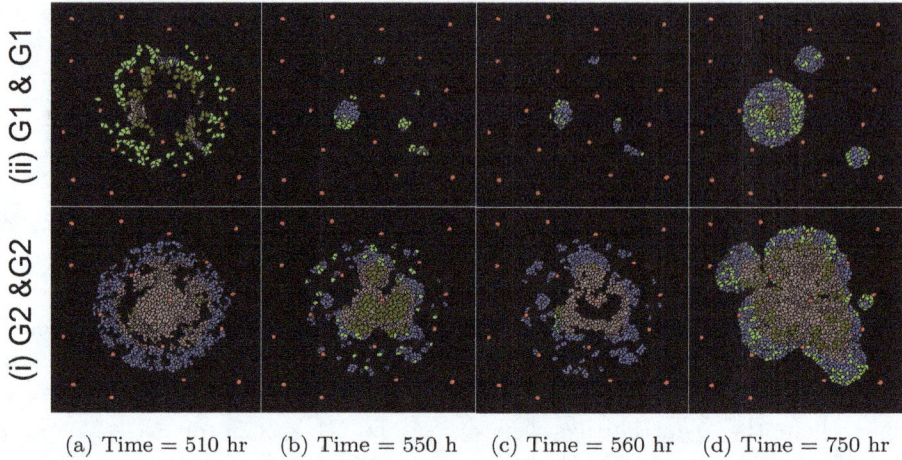

(a) Time = 510 hr   (b) Time = 550 h   (c) Time = 560 hr   (d) Time = 750 hr

Fig. 5.   Plots showing the spatial evolution of tumour cells when cell-cycle phase specific chemotherapeutic drugs are given. (i) G2 drug followed by G2 drug and (ii) G1 drug followed by G1 drug at times (a) 510 hr, (b) 550 hr, (c) 560 hr and (d) 750 hr. Adapted from Powathil et al.[6]

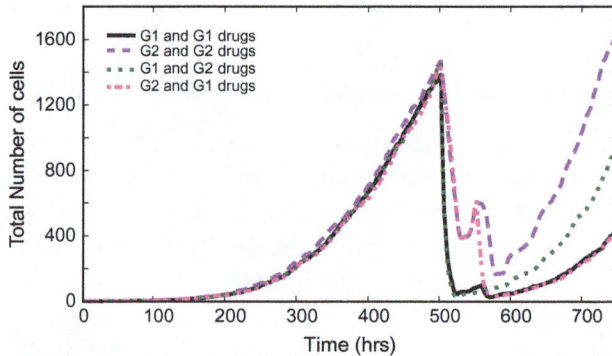

Fig. 6.   Plots comparing the total number of cells when the tumour cells are treated with two doses of cell-cycle phase specific drugs at Time = 500 hr and Time = 550 hr. Adapted from Powathil et al.[6]

and microenvironment heterogeneities play important roles in determining the precise cytotoxic effectiveness of cell-cycle phase-specific chemotherapeutic drugs.

The results of multiple combinations of cell-cycle specific chemotherapeutic drugs (Fig. 6) show that a combination of G1 specific drug followed by another G1 specific drug (Fig. 5(ii)) and G2 specific drugs and G1 specific drugs give better therapeutic outcomes than other two combinations. This is due to the presence of a higher fraction of proliferating cells in G1-phase at the time of the drug doses and increased proliferation after the initial dose. However, please note that these drug combinations need not always give the best outcome, especially if there were a higher proportion of resting cells within a growing tumour mass.[5] Hence, it is important to know the underlying spatial distribution of a growing tumour mass and the internal cellular heterogeneities present to achieve the best possible outcome.

4.2. *Heterogeneous population model: The effects of chemotherapy*

One of the common reasons for chemotherapeutic failure in cancer patients is the emergence of drug resistance in subpopulations within the growing tumour.[2] There are several reasons that contribute to this chemotherapeutic drug resistance, including multi-drug resistance to the chemotherapeutic drugs and the emergence of heterogenous subpopulations with varying responsiveness to the given drug.[33,34] Recently, it has been shown that the tumour heterogeneity caused by the cell-cycle dynamics and the variations in the cell-cycle duration can play a vital role in the chemotherapeutic sensitivity, as most of the chemotherapeutic drugs act on actively cycling cells. Several studies involving heterogenous tumour masses that contain a slowly-cycling subpopulation of tumour cells indicated that the use of traditional chemotherapeutic drugs could ultimately lead to an emerging subpopulation of drug resistant, slowly-cycling tumour cells that has the potential to repopulate the tumour mass.[33,34] Moreover, the results from recent computational studies using multiscale mathematical models have also confirmed the role of slowly-cycling tumour subpopulations in developing chemotherapeutic resistance and showed that conventional chemotherapy may sometimes result in the emergence of dominant, slowly-cycling subpopulations of tumour cells.[6]

Recently, Powathil *et al.*[6] studied the chemotherapeutic effects of anti-cancer drugs on a tumour mass that consists of two subpopulations: one with an active cell-cycle with a cell-cycle length of 25–30 hours, and a second subpopulation with slowly-cycling tumour cells. Figure 7 shows the spatio-temporal evolution of the heterogenous tumour mass with two subpopulations of tumour cells. The slowly-cycling tumour subpopulation is introduced into the previous homogenous model (Figs. 4 and 7) through random mutations that are assumed to occur after 100 mcs

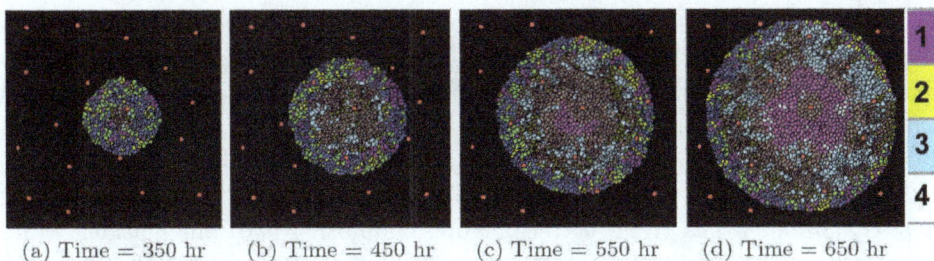

(a) Time = 350 hr   (b) Time = 450 hr   (c) Time = 550 hr   (d) Time = 650 hr

Fig. 7.   Plots showing snapshots of the simulation results of the model with two subpopulations of cancer cells at time points (a) 350 hr, (b) 450 hr, (c) 550 hr and (d) 650 hr. The colour legend given in addition to that of Fig. 4 (subpopulation 1) shows the types of the second subpopulation of tumour cells; 1 - G2 phase, 2 - G1 phase, 3 - hypoxic and 4 - resting cells. Adapted from Powathil *et al.*[6]

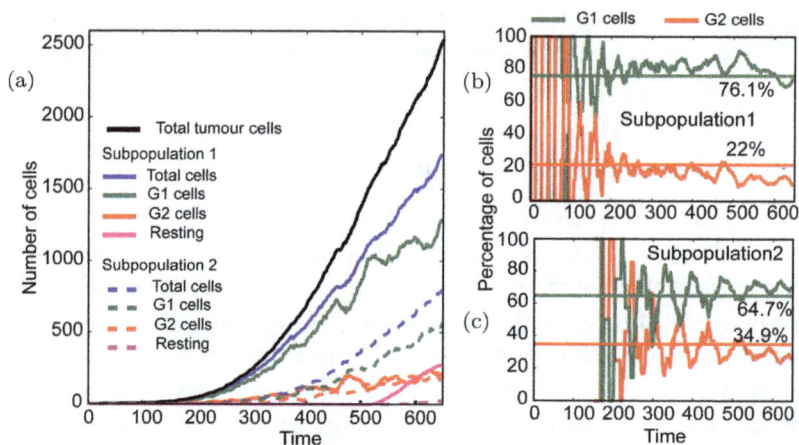

Fig. 8.   Plots showing (a) the number of cells in various phases of the cell-cycle for a heterogenous tumour with a second subpopulation of cells of slow-cycling tumour cells, (b) average percentage of cells in G1/G0 and S/G2/M phases for subpopulation 1 and (c) average percentage of cells in G1/G0 and S/G2/M phases for subpopulation 2. The lines represent the corresponding temporal average number of cells either in G1 phase or S-G2-M phases.

(hr). The quantitative results of the heterogenous tumour growth model is given in Fig. 8. Figure 8(a) compares the total number of tumour cells and the number of cells in various phases of the cell-cycle for both subpopulations 1 and 2, and Figs. 8(b) and 8(c) give the percentage of proliferating cells in subpopulations 1 and 2. The results shown in Figs. 8(b) and 8(c) indicated that the slow-cycling subpopulation has more cells in G2 phase when compared to subpopulation 1, as observed in previous experimental studies.[33,34]

The heterogenous two population tumour growth model described above can also be used to study the effects of cell-cycle phase-specific chemotherapy.[6] Two doses of cell-cycle phase-specific chemotherapeutic drugs are given at times 500 hours and 550 hours, in a similar manner to that of the homogenous case. A representative figure for the spatial evolution of cancer cells during and after the chemotherapeutic treatment with two doses of G1-phase specific drugs and G2-phase specific drugs is shown in Fig. 9. Figure 10 shows the percentage of proliferating cells in subpopulations 1 and 2 when the tumour mass is treated with each combination of cell-cycle phase specific chemotherapeutic drugs. A comparison of the total number of tumour cells and the number of cells in each subpopulations is given in Fig. 11 and it shows that combinations of G2 & G1 specific drugs and G1 & G1 specific drugs give a better outcome than other two combinations. Moreover, it can be seen from Fig. 10 that a second dose of G1-phase drug kills a majority of the cancer cells in subpopulations 1 and 2, enriching the slowly-cycling cells in subpopulation 2. These results by Powathil *et al.*[6] are in good qualitative agreement with the experimental results of Moore *et al.*[33,34] They have shown that when a heterogenous tumour mass responds positively to chemotherapeutic treatment,

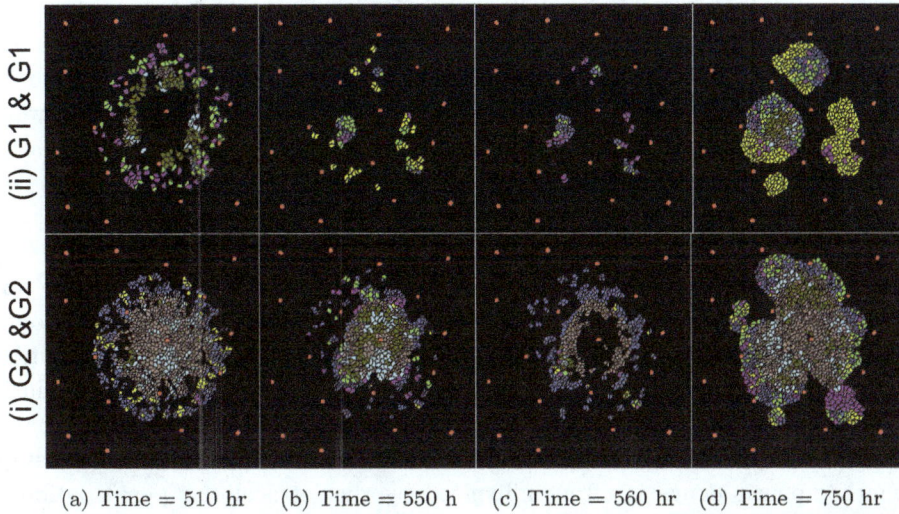

(a) Time = 510 hr   (b) Time = 550 h   (c) Time = 560 hr   (d) Time = 750 hr

Fig. 9.   Plots showing the spatial evolution of tumour cells within a two population model when cell-cycle phase specific chemotherapeutic drugs are given. (i) G2 drug followed by G2 drug and (ii) G1 drug followed by G1 drug at times (a) 510 hr, (b) 550 hr, (c) 560 hr and (d) 750 hr. Adapted from Powathil *et al.*[6]

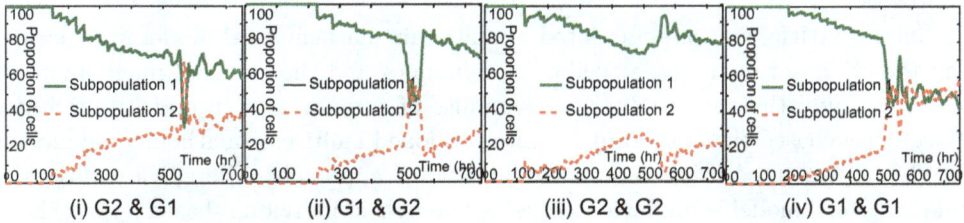

(i) G2 & G1   (ii) G1 & G2   (iii) G2 & G2   (iv) G1 & G1

Fig. 10.   Plots showing percentage of proliferating cells in subpopulations 1 and 2. Adapted from Powathil *et al.*[6]

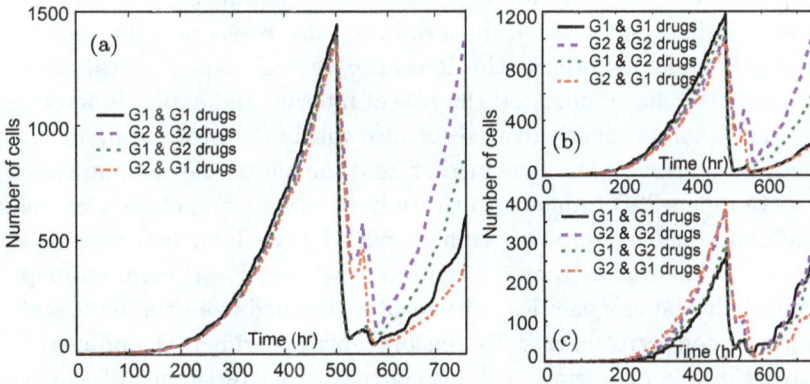

Fig. 11.   Plots showing the effects of cell-cycle specific chemotherapeutic drugs on the total number of tumour cells. (a) Total number of cells in two population model, (b) number of cells in subpopulation 1 and (c) number of cells in subpopulation 2. Adapted from Powathil *et al.*[6]

it kills a majority of the active cells, increasing the percentage of slowly-cycling tumour cells within the tumour mass (as shown in Fig. 10(iv)) and thus increasing the chance for tumour recurrence. Further extensive analysis and results of the multiscale model for a heterogenous tumour population can be found in Powathil et al.[6]

5. Conclusions

In most cases, chemotherapy is administered as a combination of multiple anticancer drugs that target various processes involved in cancer growth and progression. Combination therapy is usually used to increase the cytotoxicity and mostly targets various intracellular biochemical concentrations that are fluctuating during the cell-cycle, aiming to reduce the drug resistance due to the heterogenous nature of the tumours, with minimal toxicity. However, the efficacy of these administered chemotherapeutic drugs is often influenced by the intracellular perturbations of the cell-cycle dynamics, inducing a cell-cycle-mediated drug resistance. Hence, it is very important to study and analyse the underlying heterogeneity within a cancer cell and within a solid tumour mass due to the presence of the microenvironment and the cell-cycle position so as to design and develop more effective treatment protocols.

In this article, we have presented a multiscale mathematical model incorporating the effects of intracellular cell-cycle dynamics and the external microenvironment to study the spatio-temporal dynamics of tumour growth and its response to cell-cycle based chemotherapy.[5,6] The developed multiscale mathematical model can be implemented using various computational approaches. Here, the multiscale mathematical model is implemented using two hybrid individual-based approaches, namely: (i) a hybrid cellular automaton approach[5,9] and (ii) a hybrid cellular potts approach using the CompuCell3D framework.[6] Although each of these modelling approach is technically different, the results obtained from both multiscale models are similar and comparable, showing the robustness of the multiscale mathematical modelling approach. Recently, several experimental and clinical observations[14,33,34] have indicated the role of internal and external heterogeneities in inducing chemotherapeutic drug resistance within a growing tumour, increasing its chances of recurrence. We have further used the multiscale computational model (using the Compucell3D framework) to study the effects and efficacy of chemotherapy on a homogeneously growing tumour with intracellular heterogeneities and a heterogenous tumour growth model (with a slowly-cycling tumour subpopulation) with intracellular heterogeneities. The results obtained from the multiscale model were in very good agreement with the previous experimental findings[6,33,34] and highlighted the role of intrinsic cell-cycle-driven drug resistance of slowly-cycling tumour subpopulation in the recurrence of the tumour after therapy. Future work will consider other factors that may induce drug resistance within a growing tumour mass such as variations in cell-cycle control, anti-apoptotic proteins, multi-drug resistance through the activation of cellular pumps and increased metabolic activities[33,35,4] to study their role in tumour recurrence and analyse the various optimum delivery protocols for multiple chemotherapeutic drugs to achieve maximum therapeutic benefit.

Acknowledgments

The authors gratefully acknowledge the support of the ERC Advanced Investigator Grant 227619, M5CGS — From Mutations to Metastases: Multiscale Mathematical Modelling of Cancer Growth and Spread.

References

1. J. C. Bailar and H. L. Gornik, *N. Engl. J. Med.* **336**, 1569–1574 (1997).
2. M. A. Shah and G. K. Schwartz, *Clin. Cancer Res.* **7**, 2168–2181 (2001).
3. L. Ding, T. J. Ley, D. E. Larson, C. A. Miller, D. C. Koboldt, J. S. Welch, J. K. Ritchey, M. A. Young, T. Lamprecht, M. D. McLellan, J. F. McMichael, J. W. Wallis, C. Lu, D. Shen, C. C. Harris, D. J. Dooling, R. S. Fulton, L. L. Fulton, K. Chen, H. Schmidt, J. Kalicki-Veizer, V. J. Magrini, L. Cook, S. D. McGrath, T. L. Vickery, M. C. Wendl, S. Heath, M. A. Watson, D. C. Link, M. H. Tomasson, W. D. Shannon, J. E. Payton, S. Kulkarni, P. Westervelt, M. J. Walter, T. A. Graubert, E. R. Mardis, R. K. Wilson and J. F. DiPersio, *Nature* **481**(7382), 506–510 (2012).
4. V. Koshkin and S. N. Krylov, *PLoS ONE* **7**(7), e41368 (2012).
5. G. G. Powathil, K. E. Gordon, L. A. Hill and M. A. J. Chaplain, *J. Theor. Biol.* **308**, 1–9 (2012).
6. G. G. Powathil, M. A. J. Chaplain and M. Swat, arXiv:1407.0865 [q-bio.TO] (2014).
7. N. Goda, H. E. Ryan, B. Khadivi, W. McNulty, R. C. Rickert and R. S. Johnson, *Mol. Cell. Biol.* **23**, 359–369 (2003).
8. L. B. Gardner, Q. Li, M. S. Park, W. M. Flanagan, G. L. Semenza and C. V. Dang, *J. Biol. Chem.* **276**, 7919–7926 (2001).
9. G. G. Powathil, D. J. Adamson and M. A. J. Chaplain, *PLoS Comput. Biol.* **9**(7), e1003120 (2013).
10. J. H. Goldie and A. J. Coldman, *Cancer Treat Rep.* **63**(11–12), 1727–1733 (1979).
11. O. Lavi, M. M. Gottesman and D. Levy, *Drug Resist. Updat.* **15**(1–2), 90–97 (2012).
12. P. Vaupel, *Strahlenther Onkol* **166**, 377–386 (1990).
13. R Weinberg, *The Biology of Cancer* (Garland Science, Taylor and Francis Group, 2006).
14. G. K. Schwartz and M. A. Shah, *J. Clin. Oncol.* **23**, 9408–9421 (2005).
15. R. M. Douglas and G. G. Haddad, *J. Appl. Physiol.* **94**, 2068–2083 (2003).
16. M. D. Garrett and A. Fattaey, *Curr. Opin. Genet. Dev.* **9**, 104–111 (1999).
17. J. J. Tyson and B. Novak, *J. Theor. Biol.* **210**, 249–263 (2001).
18. B. Novak and J. J. Tyson, *Biochem. Soc. Trans.* **31**, 1526–1529 (2003).
19. B. Novak and J. J. Tyson, *J. Theor. Biol.* **230**, 563–579 (2004).
20. T. Alarcon, H. M. Byrne and P. K. Maini, *J. Theor. Biol.* **229**, 395–411 (2004).
21. A. A. Patel, E. T. Gawlinski, S. K. Lemieux and R. A. Gatenby, *J. Theor. Biol.* **213**, 315–331 (2001).
22. A. Dasu, I. Toma-Dasu and M. Karlsson, *Phys. Med. Biol.* **48**, 2829–2842 (2003).
23. G. Powathil, M. Kohandel, M. Milosevic and S. Sivaloganathan, *Comput. Math. Methods Med.* **2012**, 410602 (2012).

24. N. Goda, S. J. Dozier and R. S. Johnson, *Antioxid. Redox Signal.* **5**, 467–473 (2003).
25. A. R. Kansal, S. Torquato, G. R. Harsh IV, E. A. Chiocca and T. S. Deisboeck, *BioSystems* **55**, 119–127 (2000).
26. A. R. Anderson and M. A. Chaplain, *Bull. Math. Biol.* **60**, 857–899 (1998).
27. S. Turner and J. A. Sherratt, *J. Theor. Biol.* **216**, 85–100 (2002).
28. T. Alarcon, H. M. Byrne and P. K. Maini, *J. Theor. Biol.* **225**, 257–274 (2003).
29. B. T. Ribba, T. Alarcon, K. Marron, P. K. Maini and Z. Agur, *Lect. Notes Comput. Sci.* **3305**, 444–453 (2004).
30. P. Gerlee and A. R. Anderson, *J. Theor. Biol.* **246**, 583–603 (2007).
31. J. Moreira and A. Deutsch, *Advances in Complex Systems (ACS)* **5**(02), 247–267 (2002).
32. V. Andasari, R. T. Roper, M. H. Swat and M. A. Chaplain, *PLoS ONE* **7**(3), e33726 (2012).
33. N. Moore, J. Houghton and S. Lyle, *Stem Cells Dev.* **21**(10), 1822–1830 (2012).
34. N. Moore and S. Lyle, *J. Oncol.* **2011** (2011).
35. N. Navin, J. Kendall, J. Troge, P. Andrews, L. Rodgers, J. McIndoo, K. Cook, A. Stepansky, D. Levy, D. Esposito, L. Muthuswamy, A. Krasnitz, W. R. McCombie, J. Hicks and M. Wigler, *Nature* **472**(7341), 90–94 (2011).

Chapter 2

Physical View on the Interactions Between Cancer Cells and the Endothelial Cell Lining During Cancer Cell Transmigration and Invasion

Claudia T. Mierke

Faculty of Physics and Earth Science
Institute of Experimental Physics I
Biological Physics Division
University of Leipzig
Linnéstr. 5, 04103 Leipzig, Germany
claudia.mierke@uni-leipzig.de

There exist many reviews on the biological and biochemical interactions of cancer cells and endothelial cells during the transmigration and tissue invasion of cancer cells. For the malignant progression of cancer, the ability to metastasize is a prerequisite. In particular, this means that certain cancer cells possess the property to migrate through the endothelial lining into blood or lymph vessels, and are possibly able to transmigrate through the endothelial lining into the connective tissue and follow up their invasion path in the targeted tissue. On the molecular and biochemical level the transmigration and invasion steps are well-defined, but these signal transduction pathways are not yet clear and less understood in regards to the biophysical aspects of these processes.

To functionally characterize the malignant transformation of neoplasms and subsequently reveal the underlying pathway(s) and cellular properties, which help cancer cells to facilitate cancer progression, the biomechanical properties of cancer cells and their microenvironment come into focus in the physics-of-cancer driven view on the metastasis process of cancers. Hallmarks for cancer progression have been proposed, but they still lack the inclusion of specific biomechanical properties of cancer cells and interacting surrounding endothelial cells of blood or lymph vessels. As a cancer cell is embedded in a special environment, the mechanical properties of the extracellular matrix also cannot be neglected. Therefore, in this review it is proposed that a novel hallmark of cancer that is still elusive in classical tumor biological reviews should be included, dealing with the aspect of physics in cancer disease such as the natural selection of an aggressive (highly invasive) subtype of cancer cells displaying a certain adhesion or chemokine receptor on their cell surface.

Today, the physical aspects can be analyzed by using state-of-the-art biophysical methods. Thus, this review will present current cancer research in a different light from a physical point of view with respect to cancer cell mechanics and the special and unique role of the endothelium on cancer cell invasion.

The physical view on cancer disease may lead to novel insights into cancer disease and will help to overcome the classical views on cancer. In addition, in this review it will be discussed how physics of cancer can help to reveal and propose the functional mechanism which cancer cells use to invade connective tissue and transmigrate through the endothelium to finally metastasize.

Finally, in this review it will be demonstrated how biophysical measurements can be combined with classical analysis approaches of tumor biology. The insights into physical interactions between cancer cells, the endothelium and the microenvironment may help to answer some "old," but still important questions in cancer disease progression.

Abbreviations

MLCK : myosin light chain kinase
RLC   : regulatory light chain
EMT   : epithelial-mesenchymal transition

1. Introduction

Interactions between cancer cells and endothelial cells play an important role during the malignant progression of cancer. Malignant cancer progression involves the process of metastasis and is the worst scenario in cancer disease as it is the main cause of cancer deaths. The process of metastasis includes many steps, which follow a linear propagation. The cascade of metastasis starts with the spreading of cancer cells from the primary tumor, which migrate into the local tumor microenvironment. Moreover, these cancer cells transmigrate into blood or lymph vessels (intravasation), get transported through the vessel flow, adhere to the endothelial cell lining, grow and form a secondary tumor either in the vessel or cancer cells possibly transmigrating through the endothelial vessel lining (extravasation) into the extracellular matrix of connective tissue. These cancer cells then migrate further deeply into the targeted tissue, grow and form a secondary tumor, which means that the tumor has metastasized (Fig. 1).

Many aspects of classical tumor biology research have been investigated and thus, eight hallmarks have been postulated such as sustained proliferative signaling, evading growth suppressors, avoiding immune destruction, activation of invasion and metastasis, enabling and promoting replicative immortality, induction of angiogenesis, resistance to cell death and deregulation of cellular energetics.[1,2] However, these proposed hallmarks of cancer do not account for the physical aspects of cancer and their role in malignant cancer progression.

Among the molecules regulating cancer cell motility and invasion are cell-cell adhesion receptors such as E-cadherin-Notch signaling, cell-matrix adhesion receptors such as integrin receptors ($\alpha6\beta4, \alpha v\beta3, \alpha v\beta5, \alpha5\beta1$) and chemokine receptors such as CXCR2 and CXCR4.[3-9] All these proteins may also play a role in cancer cell–endothelial cell interactions during cancer metastasis. Despite all these current findings and even the novel approaches based on genomics and proteomics, cancer research does not fundamentally change the cancer death rates, but improves clinical diagnosis substantially in the field of cancer research regarding the classification and detailed staging of tumors, numerous marker proteins, and mapping of specific human cancer-types.

Despite these biological improvements, a main criticism remains: the expression levels of numerous genes and molecules, which are differently regulated during cancer progression, depend on the cancer disease stage. In particular, it is still

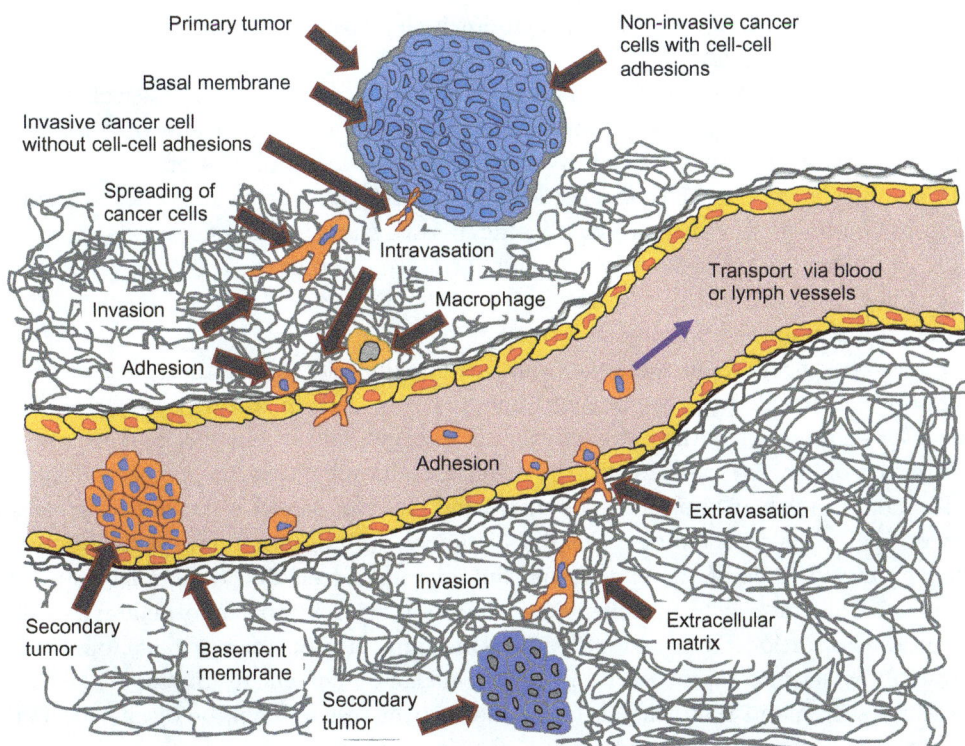

Fig. 1. The consecutive steps of the process of the metastasis formation cascade. Certain selected cancer cells can weaken their cell-cell adhesions, cross the tumor boundary including the basement membrane around the primary tumor, disseminate from the primary tumor, invade into the 3D microenvironment such as the extracellular matrix of connective tissue, transmigrate through the basement membrane of blood or lymph vessels (intravasation) possibly with the help of tumor-associated macrophages. Once entered the vessels, cancer cells are transported through the whole body and possibly transmigrate through the endothelial vessel lining (extravasation) and the basement membrane to migrate into the extracellular matrix of connective tissue and form a secondary tumor in cancer-type specific targeted organs. Another possibility is that cancer cells stick in the vessels after they adhered to the endothelial cell wall, assemble and grow inside the vessel as a secondary tumor.

not fully understood how they regulate cancer progression. A reason may be that these genomic and proteomic based methods do not account for the localization of molecules in special compartments such as lipid rafts,[10] their activation or assembly state, their life-time, turn-over, modification and recycling rate.[11–15]

As the whole complexity of malignant cancer progression seems not to be covered by the genomic or proteomic based methods, biophysical methods are promised to obtain more insights into malignant cancer disease progression. In more detail, classical physical approaches will be adopted to complex soft matter such as cancer cells and novel biophysical methods will be developed in order to apply them to cancer research. These novel physical approaches have changed the direction of recent cancer research and have broken down the classical view on cancer disease during the last decade.

In particular, a novel hallmark adding the aspect of physics to classical cancer research is that the primary tumor and the tumor microenvironment alter the survival conditions (such as cell-division) and cellular properties of a certain set of cancer cells, which subsequently lead to the selection of an aggressive (highly invasive) subtype of cancer cells. This aggressive subtype of cancer cells may be able to reduce cell-cell adhesions to neighboring cells, cross the tumor boundary of the primary tumor, invade into the extracellular matrix scaffold and transmigrate through the endothelial lining of vessels.

This review will focus mainly on our work of the selection of an aggressive subtype of cancer cells expressing high amounts of a certain integrin or chemokine receptor on their cell surface. In addition, it will highlight the impact of mechanical properties of invasive cancer cells, the mechanical properties of the cancer microenvironment, and their impact on the migration mode of cancer cells. Moreover, this review will focus on the dimensionality of cancer cell migration, on the novel invasion promoting role of the endothelial lining from vessels and on the mechanical properties of cancer cells and their possible modulation by the endothelium as well as on the alterations of the endothelium's mechanical properties. Finally, our findings will be set into the broad range of knowledge on current results in the field of cancer research and their impact on further cancer research will be discussed. The next two chapters will discuss how the selection process of an aggressive cancer cell phenotype can take place.

2. Selection of Aggressive Cancer Cell Subtypes Expressing a Certain Cell-Matrix Receptor

The selection of an aggressive phenotype of cancer cells in a primary tumor seems to be the onset of the malignant cancer progression cascade. Although the process of metastasis occurs rarely among the numerous cancer cells within a primary tumor, once started the metastasis has a worse prognosis for the patient. In principle one single cancer cell may be sufficient to cause malignant cancer progression. However, a few of these cancer cells undergo a transition or selection process in order to be able to follow the step-by-step series of the metastatic progress. How this selection process works is still elusive. We suggest that alterations of the mechanical properties may play a pronounced role in selecting aggressive and highly invasive cancer cells. The following gives an example of how such a selection process may work and when it may start during cancer disease progression.

The process of cancer metastasis is a complex process that includes sequential steps and is responsible for over 90% of cancer-related deaths. When focusing on the onset of metastasis, cancer cells spread from the primary tumor, cross the tumor boundaries such as the basement membrane and tissue compartments, migrate or flow through vastly different microenvironments, including the tumor stroma, the blood vessel endothelium, the vascular system and the tissue at a secondary target site,[16,17] where the probability of metastasis occuring is increased compared to other non-targeted sites (Fig. 1). What role the mechanical properties of the home tissue of the primary tumor and the secondary target tissue does play is still under investigation and discussion. Currently, we suggest that similar mechanical properties of the two tissue types may enhance the ability of cancer cells to build a secondary tumor at this particular site. Until now, it is not known what determines the specificity of target tissues for cancer metastasis for a certain cancer type. Together with enhanced cellular invasiveness, specific cellular morphology, a certain cytoskeletal architecture and certain biomechanical properties, cancer cells can restructure and thus even adapt their microenvironment[18-20] to promote tumor progression and finally to metastasize in targeted organs. These findings may suggest that cancer cells need a special microenvironment, where they are able to alter biochemical and mechanical properties. As we have analyzed the expression profile of highly invasive cancer cell types and non-invasive cancer cell types, when co-cultured with endothelial cells for 16 hours compared to monocultured cancer cell types, our findings are that the $\alpha5\beta1$ integrin and the $\alpha v\beta3$ integrin expression are both increased in co-cultured highly invasive cancer cells compared to monocultured cancer cells, but not in co-cultured non-invasive cancer cells compared to monocultured cancer cells.[3]

These results indicate that the afore-mentioned two integrins may play a role in cancer cell transendothelial migration and invasion. Therefore, we selected subcell lines from parental cell lines such as MDA-MB-231 (breast), T24 (bladder), 786-O (kidney) and A375 (skin) using a flow cytometer, which express high, intermediate and low amounts of $\alpha5\beta1$ or $\alpha v\beta3$ integrins on their cell surface. Indeed, these subcell lines showed altered invasive properties due to integrin expression levels.[21,22] In particular, the metastatic signal transduction cascade is supposed to be triggered by the proliferation of the primary tumor, genetic alterations of primary cancer cells, activation of signaling pathways and the selection of an aggressive cancer cell subtype. This aggressive cancer cell subtype should be able to weaken the cell-cell adhesions and be able to cross the tumor boundaries such as other cancer cells within the primary tumor and the basement membrane. These initial steps may further promote the invasiveness of this special subtype of cancer cells including their ability to form protrusions, intravasate into blood or lymph vessels and finally, to metastasize. Despite the fact that many steps are not understood in detail, however, how the detachment of a certain subtype of cancer cells from the primary tumor epithelium occurs and how this subtype invades the underlying tumor stroma is able to be studied at the biomechanical, cellular and molecular levels. In summary, this process can be described by the well-known epithelial-mesenchymal transition (EMT), which was initially defined in the process of embryogenesis.[23] The role of the EMT in cancer metastasis has been extensively studied[24,25] and is still controversially discussed. Agreement has been reached about the key components of the EMT that are E-cadherin, Notch receptors (both cell-cell adhesion molecules), matrix-metalloproteinases (that degrade the extracellular matrix of connective tissue) and cytokeratins, which can lead to enormous changes in the physical and mechanical properties of cancer cells. In particular, the reduction of cell-cell adhesion between neighboring epithelial cancer cells and morphological alterations of the cell shape from cuboidal epithelial to fibroblastoid mesenchymal[26] as well as in their aspect ratio.[21] These alterations may cause the loss of intercellular adhesions, which initially hold the primary tumor together and may now serve as a selection pressure that is applied to select for an aggressive and highly invasive possibly mesenchymal cancer cell subtype, which can invade the tumor microenvironment.[21,25]

However, it has still not been fully investigated what regulates the molecular and physical mechanisms that enhance the invasiveness of these special subtype of aggressive cancer cells and enable them to migrate into the tumor microenvironment. A question arises whether these subtypes of cancer cells possess specific stroma adhesion receptors in order to migrate through antigens such as CD44, which can bind to hyaluronic acid (increased in the surrounding tumor microenvironment).

Not only can the stroma alter the embedded tumor by accumulation, binding or production of substances (by stroma cells), cancer cells may also alter their microenvironment drastically in order to migrate into it and invade. The movement of cancer cells in a 3D microenvironment also depends on the enzymatic digestion of the extracellular matrix or by sheddases cutting cell-matrix adhesion receptors on the cell surface of the cancer cells such as ADAM-10 and ADAM-17 (see below). This enzymatic mode of cell invasion may follow other constrictions compared to the non-enzyme-driven cell invasion.[20]

However, the integrin-dependent mode functions in the presence and absence of enzymatic digestion. Recently, we found that $\alpha5\beta1$ integrin expression facilitates cancer cell invasion into the extracellular matrix and even enhances the transmigration through endothelial cell monolayers grown to confluent monolayers on top of 3D extracellular matrices.[4,21] The latter finding was surprising and unexpected as the endothelium was always presented as a strong "passive" barrier for cancer cell invasion. In contrast, we reported for the first time that the endothelium can act as an enhancer or inducer of cancer cell invasiveness into 3D extracellular matrices.[21] In addition, also $\alpha v\beta3^{\text{high}}$ integrin expressing cancer cells showed increased invasiveness into 3D extracellular matrices compared to $\alpha v\beta3^{\text{low}}$ cells.[22] What about other integrins and their effect on cancer cell invasion?

Finally, other integrins such as collagen binding integrins such as $\alpha1, \alpha2, \alpha10$ or $\alpha11$ have to be analyzed for their potential in regulating cancer cell invasion. Besides the cell-matrix adhesion receptors such as integrins, other cell surface receptors such as chemokines may play a role in cellular invasiveness. Thus, another approach for the selection of a highly aggressive subtype of cancer cells is discussed in the next chapter below.

3. Selection of Aggressive Cancer Subtypes Expressing a Chemokine Receptor

In the chapter above we have discussed the selection of cancer cell lines expressing high and low amounts of integrins. Now we will discuss the selection process of cancer cells expressing high and low amounts of the chemokine receptor CXCR2 and their subsequent ability to invade and transmigrate. Simultaneously, as we determined the expression pattern of the highly invasive and non-invasive cancer cells in co-culture with endothelial cells and in monoculture, we also analyzed the endothelial cells in co-culture and monoculture with highly invasive and non-invasive cancer cells. We report that chemokines such as Gro-β (CXCL2) and interleukin-8 (CXCL8) are increasingly expressed when primary microvascular endothelial cells are co-cultured with highly invasive cancer cells compared to monocultured endothelial cells, but not when co-cultured with non-invasive cancer cells.[4] As the receptor for these two chemokines is CXCR2, which is increasingly expressed on highly invasive cancer cells compared to non-invasive cancer cells, we isolated subclones from parental breast cancer cells (MDA-MB-231) with high and low CXCR2 expressions on their cell surface (Fig. 2). These CXCR2^{high} subcell lines showed increased invasiveness compared to the CXCR2^{low} subcell lines in 3D extracellular matrix and even the transmigration through an endothelial cell monolayer was increased in terms of numbers and invasion depths. Moreover, these CXCR2^{high} and CXCR2^{low} subcell lines possess different cellular stiffness and altered contractile forces transmission and generation that may explain their different invasiveness. How mechanical properties of cancer cells account for increased invasiveness will be discussed in the following chapter.

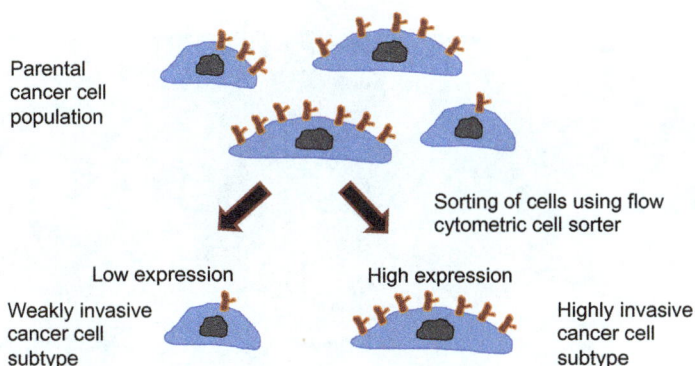

Fig. 2. Schematic drawing of the selection process of highly and weakly invasive cancer cells from a parental human cancer cell line. A cancer cell line shows weakly and highly expression of the CXCR2 chemokine receptor on its cell surface. Thus, the cells are selected as single clones showing either weakly or highly expression of the CXCR2 receptor. These cells grow to cell lines and are called subgroup cancer cell lines.

4. Mechanical Properties of Cancer Cells Facilitate Invasion

It is still not known what the prerequisites for the cell motility in 2D and 3D systems are. However, the ability of cancer cells (of epithelial origin) to migrate into 3D connective tissue does not depend on a single parameter. Instead, it rather depends on certain mechanical parameters, which regulate the migration velocity of cancer cells through a dense 3D extracellular matrix (pore size around $2\,\mu$m). Among these parameters are certain properties such as: (a) cell adhesion and de-adhesion dynamics (turn-over of focal adhesions, adhesion strength), (b) cytoskeletal remodeling dynamics, cell-fluidity and cell-stiffness, (c) matrix remodeling by secretion of extracellular matrix proteins and digestion through matrix degrading enzymes and (d) the generation and transmission of protrusive or contractile forces.[4,27–30] Each of these parameters cannot be treated as a single one, they must be related to the others to see how strong the particular effect on cell invasion and transendothelial migration is (Fig. 3). Thus, the balance between these parameters is crucial for the efficiency of cancer cell invasion, the speed of invasion and invasion depths in 3D extracellular matrices.[31] These parameters can vary depending on the cancer cell type and shift towards a single parameter, but still these parameters are all important for determining cancer cell invasiveness and invasion efficiency except for cancer cell types which are not of epithelial origin. In this case, these parameters vary a lot and some may even play no role.

For cancer cells of epithelial origin, a disruption of the balance between these parameters leads to a drastic change of the invasion mode from epithelial to mesenchymal, or even to amoeboid with or without traction forces. Whether this transition is always full and holds true for all types of cancer cells needs to be further investigated. Taken together, the invasion strategy is determined by these

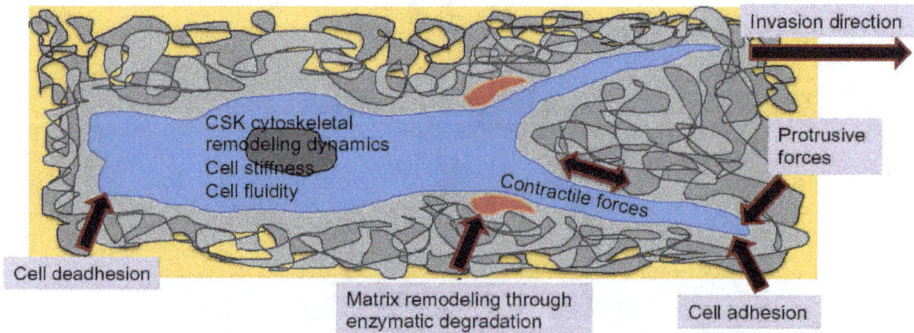

Fig. 3.   Schematic image of a cancer cells invading a 3D matrix and the parameters regulating cancer cell invasion into dense 3D matrix scaffolds. The ability to invade through the extracellular matrix of connective tissue is regulated by a balance between at least four biomechanical processes: (i) contractile force generation and transmission as well as protrusive forces, (ii) transmission of contractile forces via cell-matrix adhesions (adhesion strength, de-adhesion), (iii) cytoskeletal (CSK) remodeling dynamics, cell fluidity, and stiffness, and (iv) enzymatic matrix degradation.

biomechanical and biochemical parameters. How the balance of these parameters is regulated by cancer cells (of epithelial origin) and what role the tumor microenvironment regarding growth factors, cytokines, chemokines, matrix-protein composition, structure and concentration, and matrix mechanical stiffness plays is still not fully clear and thus under intensive investigation.[17]

There are two different mechanisms that are currently presented and discussed: The first mechanism is the degradation of the dense 3D extracellular matrix through the secretion of matrix metalloproteinases (MMPs) in order to facilitate cancer cell invasion.[20,32,33] The second mechanism is the cutting of cell-cell adhesion molecules such as NOTCH receptors, ephrins or E-cadherins from the cell surface of cancer cells by sheddases such as the secretases ADAM-10 and ADAM-17.[34–40] This reduction of cell-cell adhesions may then facilitate signal transduction leading to nuclear translocation (together with a transcription factor) of cell-cell adhesion proteins such as β-catenin and induction of gene expression, and finally induction of cancer cell invasiveness. Furthermore, the sheddase ADAM-17 can also cleave pro-TNF-α exposed on the cell surface of cancer cells to release TNF-α into the microenvironment.[41] This may then activate nearby endothelial cells, which are subsequently stimulated to facilitate cancer cell transmigration. Besides the enzymatic degradation of shedding of membrane receptors, another parameter regulating the invasion speed and the invasion depth of cancer cells in dense 3D extracellular matrices is the physical property of cancer cells to generate and transmit contractile forces.[4,42] Recently, biophysical methods for measuring contractile forces in 3D collagen or fibrin matrices have been presented.[43–46] In more detail, the invasion of cancer cells can be analyzed by taking z-stack images using a confocal scanning microscope. In some cases, matrix embedded beads may serve as markers as well as be part of the collagen fiber structure. The tracking of collagen fibers is more sophisticated, but also more reliable. Using collagen fibers as markers, the effect of marker (bead) phagocytosis is vanished as only a minor number of collagen fibers are digested and internalized compared to embedded beads in close neighborhood to invasive cancer cells. In addition, bead internalization affects the mechanical properties of cancer cells such as stiffness and subsequently, it reduces the invasiveness of highly invasive cancer cells into dense 3D extracellular matrices.[47]

As integrins were up-regulated in invasive cancer cells, we focused on them and supposed that they were possibly key players for the regulation of cancer cell invasion into dense 3D matrices as well as their cytoplasmatically associated focal adhesion molecules such as vinculin and focal adhesion kinase (FAK). In particular, integrins are important for facilitating the invasiveness of cancer cells because they regulate cell-matrix adhesion and de-adhesion, adhesion strength, transmission and generation of contractile forces through outside in (via ligand-binding) or inside-out (via internal signaling through growth factors etc.), stimulation of integrin-receptors (clustering) as well as cytoskeletal remodeling dynamics.[48,49] The integrins are a family of cell-matrix adhesion receptors and consist of two non-covalently linked α (18 different ones) and β (8 different ones) subunits. Taken together, integrins facilitate transmembrane connections between the actomyosin cytoskeleton of the cells and their microenvironment such as an extracellular matrix fiber scaffold. In more detail, the connection between the cell's cytoskeleton and the microenvironment seems to be facilitated by the focal adhesion protein vinculin which can act as a mechano-coupling and mechano-regulating protein.[50–52] The focal adhesions of cancer cells possess two main functions in cell invasiveness: Firstly, they transmit contractile forces to the microenvironment. Secondly, they mediate the connection of cancer cells to its external substrate in order to withstand de-adhesion and apoptosis.[31,53,54] In addition, the composition of focal adhesions may vary due to the integrin type and the integrin activation state. Moreover, the signaling through α integrin subunits is not uniform among the different α integrin subunits: for example, the activation of $\alpha 4$ and $\alpha 9$ integrin subunits diminishes the cell spreading, whereas the activation of $\alpha 1$, $\alpha 3$, $\alpha 5$, $\alpha 6$, αv and αIIb integrin subunits increases the cell spreading.[55–59] Due to the integrin type and integrin activation state, the functional role of vinculin in focal adhesions may also be altered. Moreover, integrins regulate the function of other integrins: the $\alpha 5 \beta 1$ integrin has been reported to regulate the facilitated adhesions as well as the motility on extracellular matrices.[60] Thus, we also investigated the effect of $\alpha v \beta 3$ highly and lowly expressing cells on cancer cell invasion into dense 3D matrices. Indeed, we found that the $\alpha v \beta 3^{\text{high}}$ cells invaded more numerously and deeply into the 3D matrices compared to $\alpha v \beta 3^{\text{low}}$ cells.[22] In addition to the impact of the mechanical properties on cancer cell functions as such as cellular motility, it is explained and discussed how cancer cells alter their microenvironment in the next chapter.

5. Cancer Cells Alter Their Microenvironment

As we have discussed above, the mechanical properties of cancer cells have a pronounced impact on their invasiveness. It has been supposed that the mechanical properties of cancer cells are determined or regulated by their microenvironment.[61] Besides the mechanical properties of cancer cells, the microenvironmental mechanics can also be altered by transmigrating cancer cells. In particular, the invasive cancer cells can express MMPs on their surface such as MT1-MMP, which promote the digestion of the laminin- and collagen IV-rich basement membrane[62] and subsequently, the digestion of the extracellular matrix tissue microenvironment leads to major restructuring of the 3D extracellular matrix, either local and global by secreting additional matrix components in the 3D matrix scaffold (Fig. 3).[21] When invasive cancer cells pass the tumor stroma, they may sense the complex composition of the connective tissue scaffold consisting of collagen type I and fibronectin through their particular cell-matrix receptors.[63] For example, in close neighborhood to the primary tumor site the connective tissue scaffold has been reported to be stiffer compared to the matrix scaffold of normal healthy connective tissue due to elevated levels of collagen[64] and additionally, due to elevated crosslinking of collagen fibers by lysyl-oxidase through cancer associated fibroblasts.[65] This also enhances the stability of collagen fiber networks and increases their resistance to matrix degradation. In more detail, the crosslinking of collagen fibers triggers the outside-in signaling of the cell-matrix receptors, the integrins, as well as the binding of certain ligand to these receptors.[66] Moreover these dramatic alterations of the extracellular matrix's structure and mechanical properties may induce and enhance the proliferation of cancer cells, increase cancer cell invasion and represent possibly a positive feedback-loop.[64] It is still elusive whether the effect of extracellular matrix stiffening of mammary tumors is transferable to other solid epithelial originated tumors. In more detail, how the stiffening process of the extracellular matrix is regulated and evoked is still not yet clear, in particular, what regulates the mechanical stiffening on the molecular level is under investigation. The knowledge of these molecular and biochemical regulation processes may help to understand how the whole process of cancer cell invasion starts and hence how it could be inhibited. Recently, we reported that the $\alpha5\beta1$ integrin facilitates cancer cell invasion into the extracellular matrix, which needs to have a certain mechanical stiffness.[21,47] Moreover, the $\alpha5\beta1$ integrin facilitates cancer cell invasion even through endothelial cell monolayers grown to confluent layers on top of 3D extracellular matrices without disrupting, destroying or remodeling of the endothelial cell monolayer.[4,21] However, it is still not yet shown what role sheddases such as ADAM-8, ADAM-10 or ADAM-17 play and whether they enhance or decrease cancer cell invasion and transendothelial migration. Taken together, this is still elusive and has to be determined in future studies. Not only do cancer cells alter the physical properties and the structure of their microenvironment, this interaction between cancer cells and their microenvironment is not restricted to cancer cells acting on the microenvironment, thus in addition the microenvironment has a broad impact on the cancer cell's properties and functions as described and discussed in the next chapter.

6. The Microenvironment Alters the Biochemical and Mechanical Properties of Cancer Cells

As the microenvironments around primary tumors are altered by the primary tumor itself and by invasive cancer cells, the mechanical properties of cancer cells may also be altered, caused by their microenvironment. The mechanical properties of tumor microenvironments are altered compared to normal healthy tissue.[67] Thus, the tumor microenvironments seem to be a key element for the regulation of cancer cell motility through 3D connective tissue and transendothelial migration through blood or lymph vessels. In line with this, we and others have reported that the microenvironment of tumors is no passive compartment. Rather, it regulates the progression of cancer disease by behaving as an active element, which is important in regulating its mechanical properties in order to provide malignant tumor progression.[68,21] In particular, the tumor microenvironment has been regarded as highly critical for all steps of the cancer metastasis process. In particular, we have demonstrated for the first time that the endothelial microenvironment of a tumor is an active element for enhancing and initiating the invasion of certain cancer cells into dense 3D extracellular matrices, where the pore size is smaller than the cancer cell's diameter.[19,30,69]

In addition, external physical properties of the tumor microenvironment can be of geometric origin such as pore size, crosslinks, concentration and composition between collagen fibers of the 3D extracellular matrix or mechanical origin such as the matrix's stiffness, which may affect the ability of cancer cells to invade into the matrices.[67,71,72] These parameters cannot be seen as independent or unrelated, they are indeed connected and cannot always be varied independently. In more detail, the structure of the tumor microenvironment is determined by matrix stiffness, pore size, connection points or crosslinking proteins, extracellular matrix fiber network composition and concentration, fiber thickness, bending and orientation of the fibers.[73-75] These parameters are not separately modulated and are not independent of each other, instead, they are rather related. In particular, for example the gel's elastic module depends on the density of crosslinks and the rigidity of the polymer chains or fibers (depending on the persistence length). Moreover, even the matrix's stiffness and the pore size are affected if the composition of the collagen types is varied. If collagen V is taken, the pore-size and the fibril diameter are supposed to be the same as if collagen type I is taken, but the matrix stiffness can be altered by varying the ratio between collagen types I and V.[76] The cancer cells and the primary tumor can alter all these parameters of the surrounding tumor microenvironment.[77,78] In addition, tumor-associated cells such as endothelial cells, macrophages or fibroblasts are able to alter or adjust these properties of matrices in order to adapt the external matrix to an optimal substrate for a single or collective invasion of cancer cells.[79,80]

When cancer cells are too stiff or too soft, however, they can probably not alter highly crosslinked collagen fibers or bundles of the extracellular matrix scaffolding through applying a force in order to enlarge the pores or bend the fibers and thus increase their migration efficiently.[64,81] Moreover, even the ability of cancer cells to degrade the extracellular matrix by MMPs may be altered and hence, is not sufficient to provide cancer cell invasion in certain circumstances. In summary, the mechanical properties of a primary tumor's local microenvironment seems to play an important role in mediating the invasive and aggressive properties of cancer cells. Our understanding of the mechanical properties has been increased, but still many important questions remain. For example, does the matrix stiffness of the local tumor microenvironment facilitate the selection of an aggressive and highly invasive cancer cell sub-type? In particular, does the selection of this invasive cancer cell subtype underlie similar principles as for the differentiation of mesenchymal stem cells into distinct lineages? The latter question is partly answered as the differentiation of stem cells has been shown to be consistent with differences in tissue compliance.[68] What role does the mechanical properties of the matrices play in facilitating cancer cell invasion and transmigration? Moreover do the invasive cancer cells form podosomes on 2D matrices or invadosomes in a 3D microenvironment?

How many cancer cell types form podosomes or invadosomes? Do cancer cells use protrusive force-dependent or blebbing surface tension-dependent invasion modes? How is the formation of these structures regulated by the proteolytic digestion of the extracellular matrix?

Taken together, the physical and mechanical interactions between a cancer cell and its extracellular matrix (a collagen-rich scaffold on which the primary tumor grows) plays a key role in allowing cancer cells to migrate from a tumor to nearby tissues by crossing the tumor boundary as well as compartment boundaries, and if the primary tumor is strongly malign, cancer cells interact with tumor blood or lymph vessels. During the steps of intravasation and extravasation, invasive cancer cells may undergo large elastic deformations to penetrate endothelial cell-cell contacts or even to migrate through a living endothelial cell lining vessel walls, which then still stays intact after the transmigration step and reassembles to a closed endothelial monolayer after the transmigration event of a cancer cell. How this is regulated is still not yet known, but it is currently under investigation. Once having entered the vascular system, a cancer cell has to deal with the vessel flow (applying shear stress on the circulating and adhering cancer cells), which may impact the cancer cell's migration velocity and adhesion strength to endothelial vessel walls. All this can also influence the binding efficiency of the cancer cells (within the vessels) to endothelial cells and may in particular determine the translocation of sites where a secondary tumor will be formed and finally grow. In summary, the knowledge of how physical interactions and mechanical forces are related to biochemical changes will help to get novel and important insights into the progression of cancer and hence form the fundamental basis for new therapeutic approaches.

In the past, it has been found that the mechanical phenotype of cancer cells growing on soft or rigid substrates was altered.[82] Moreover, the ability to build up a secondary tumor in the relatively soft tissue lung could be explained by the ability of these cancer cells to grow on soft matrices.[82] Thus, mechanical properties of the extracellular matrix play a role in regulating the invasive properties of cancer cells. During the malignant progression of cancer the rigidity of the extracellular matrix material increases,[83,84] which leads then to the accumulation of dense and crosslinked collagen fiber matrices around the primary tumor.[85] This altered cancer microenvironment may serve as a signal for cancer cells of the primary tumor to increase their invasive properties in order to be able to migrate out of the primary tumor directly into the tumor microenvironment. Which invasion or migration mode will these cancer cells use? How do the physical properties of invasive cancer cells and their surrounding microenvironment affect and determine the migration mode is discussed in the following chapter.

7. Do Physical Properties Determine the Migration Mode?

The answer to this question is properly that physical properties can determine the migration mode together with biochemical properties that can also alter the invasion mode. An example is the inhibition of matrix degrading enzymes such as matrix-metalloproteinases in carcinoma and fibrosarcoma cells using a special cocktail of inhibitory enzymes that lead to a switch of the migration mode from a predominantly integrin-based mesenchymal motility to a faster amoeboid migration mode.[33,86] Another example for a regulation of the invasion mode in the opposite direction is that an increase in contractility by using calyculin A converts a former more amoeboid invasion mode to a slightly mesenchymal invasion mode as shown for cancer cells expressing low amounts of the $\alpha 5\beta 1$ integrin.[21] The motility of cancer cells in uncrosslinked (pepsin extracted collagen type I gels) can be amoeboid with and without contractile forces,[18,42] whereas in crosslinked collagen gels the motility of cancer cells still requires MMPs such as MT1-MMP (MMP14).[63,87,88] Taken together, if the MMP function directly depends on the collagen matrix microstructure such as the collagen concentration and crosslinking density, the MMP inhibition would solely be effective in diminishing cancer cell motility in highly crosslinked and highly concentrated regions of the extracellular matrix, whereas MMPs would be totally ineffective for less crosslinked and low density regions of the matrix. In addition, the external application of mechanical forces can enhance the MT1-MMP-driven proteolysis of the extracellular matrix.[89] This finding leads to the hypothesis that there is a signal transduction-mediated connection between the forces sensing and the secretion of proteolytic enzymes. Finally, to prove this hypothesis, additional experiments are necessary. In particular, there is evidence that the physical properties of the tumor microenvironment are crucial in tumor initiation, progression and metastasis through a functional connection between physical forces and biochemical signal processes. However, many classical migration assays such as wound healing assay ignored the effect of the dimensionality on cell motility and that a 3D microenvironment includes more constraints compared to a 2D microenvironment. Therefore, this aspect is discussed in more detail in the following section.

8. Cancer Cell Motility in Three Dimensions

Previously, many physical and molecular mechanisms determining the motility speed of normal and cancer cells have been studied *in vitro* assays using two-dimensional (2D) substrates.[83–85] These assays have been called wound healing assays, where a scratch between the monolayer of a cell population re-induces their migration into the free space (gap) or inserts are used to culture cells in a monolayer until they reach the borders of these inserts; by removing the inserts, a gap is left where the cell can move in. The latter assay may be more suitable for the cells, as they are not hurt by the cut through the monolayer compared to the original classic scratch assay and additionally, these cells formed a real border when reaching the inserts.

Despite these developments, we have reported that the dimensionality of the cell-culture system used to study cell invasion is crucial for the mode of cellular migration.[21,52] The 3D microenvironment of the extracellular matrix *in vivo* is characterized by certain features such as the pore size, connection points and fiber or bundle orientation, all of which cannot be found in extracellular matrix protein-coated 2D substrates.[87] However, the underlying principle of 2D and 3D migration may be different. In more detail, several features seem to be crucial for 2D motility such as focal adhesions, stress fibers, broad lamellipodia, filopodial protrusions at the leading edge and apical polarization (Fig. 4). These 2D motility features are pronouncedly reduced in size and play no role for invasive cancer cells migrating through a 3D extracellular matrix.[31,90–93]

Recently, we have shown that the motility on 2D and 3D substrates could be quite different. In particular, the expression of the mechano-coupling and mechano-regulatory protein vinculin is important for the regulation of cellular motility in 3D, but plays no role in 2D cell motility as vinculin knock-out cells migrate faster on a 2D substrate compared to vinculin-expressing (wild-type) cells.[52] Additionally, recently it has been suggested that focal adhesions, composed of clustered integrins which structurally and dynamically couple the cellular actin-myosin cytoskeleton to the extracellular matrix proteins on 2D substrates, are altered when these cells are inside a 3D extracellular matrix.[90] Our hypothesis is that other focal adhesion proteins may behave similar to vinculin. However, there could also be differences to focal adhesion proteins that do not act as mechano-coupling proteins.

Another explanation for the differences in cellular motility on 2D substrates and in 3D matrices could give the appearance of collagen fibers in 3D extracellular matrices. Thus, the extracellular matrix may support the dynamically clustering of integrins, with sizes in the order a few nanometers and lifetimes of a few seconds, that seem to be necessary for cell invasion through a 3D microenvironment (Fig. 4).

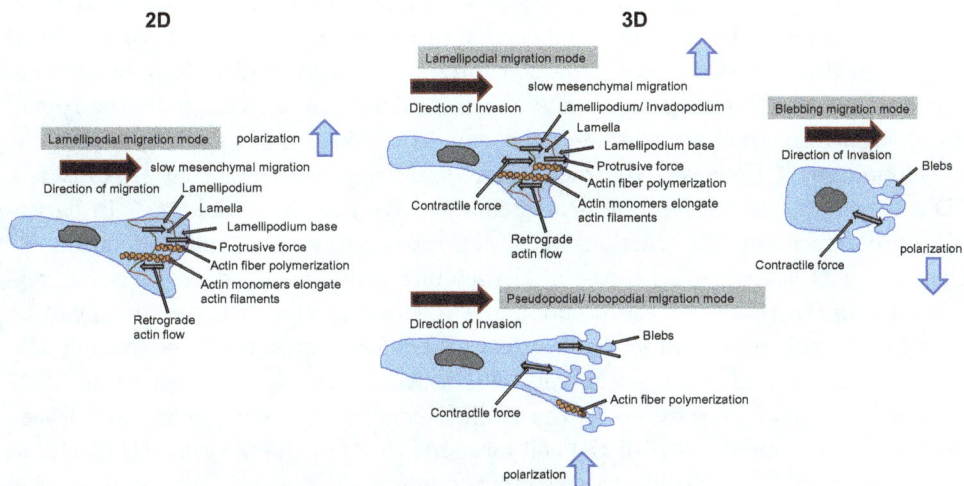

Fig. 4. Schematic drawing of the different migration/invasion modes of cancer cells due to the dimensionality of the substrate for epithelial derived cancer cell types.

In particular, cells *in vivo* can initiate the bundling of collagen fibers by the generation of contractile forces of cellular protrusions such as filopodia and invadopodia.[94] Moreover, the collagen bundles increase the surface area available for the cells that in turn induces the assembly of even larger focal adhesions.[95] It still remains elusive whether cancer cells are able to from similar focal adhesions and build up stress fiber in a 3D extracellular matrix. How does the mechanical properties of the 3D extracellular matrix modulate the focal adhesion assembly and the actin stress fiber formation? Stress fibers, containing bundled actin filaments, play a prominent role in 2D cell motility systems, where they transmit the contractile forces required for the regulation of de-adhesion of the rear of a cell from the 2D substratum and the establishment of the actin flow at the leading edge of the migrating cell.[82,96] In contrast to 2D systems, cells possess less stress fibers inside a 3D extracellular matrix compared to an extracellular matrix protein coated 2D surface. However, a quantitative analysis is still missing. In more detail, these 3D stress fibers are either localized to the cell cortex (called cortical action network) or radiate from the nucleus towards the plasma membrane to form pseudopodial protrusions.[43] In particular, inhibition experiments blocking the actomyosin contractility are often less effective in 3D cell motility systems compared to 2D cell motility systems,[43] which raises the question whether the stress fibers are dependent on dimensionality of the motility system.[96,97] In contrast, we have recently reported that the 3D cell invasion of certain highly invasive cancer cells lines could be inhibited by using inhibitors reducing the actomyosin-dependent contractility such as the myosin light chain kinase inhibitor (ML-7), the Rho-kinase inhibitor (Y27632), or latrunculin A (actin-polymerizing inhibitor).[19,21,30] In addition, it has been shown that also the apical polarization of the cells in 2D culture plays a role in cell migration, because when the polarization is reduced, the number of focal adhesions and stress fibers is significantly decreased, the functional role of focal adhesion proteins such as vinculin or FAK is fundamentally altered and certain proteins such as α-actinin or myosin-II are enriched in stress fibers.[98] Traction force microscopy results lead to the suggestion that in a 2D migration system, a lamellipodium actively pulls the rest of a cell through nascent focal adhesions newly established at the leading edge of the lamellipodium.[99] However, 3D traction force microscopy reveals that cells inside a 3D extracellular matrix pull on nearby fibers of the matrix scaffold.[43,46] In particular, pronounced matrix tractions occur close to active pseudopodial protrusions[43] that pull with nearly equal forces at the leading and trailing edges of the migrating cell. As the release of the pseudopodia towards matrix collagen fibers can be asymmetric, this may then lead to a structural defect within the fiber scaffold normally at the rear of a migrating cell. Due to these results, it seems to be likely that pseudopodial protrusions at the trailing edge of migrating cells are released first and thus, pull the rear of the cell forwards through the 3D extracellular fiber matrix scaffold[100,101] leading to persistent guided motion similar to the migration through a tunnel.

In contrast, the motion of cells in 2D is less persistent as this migration does not need to form a tunnel through which the cells walk through,[102] the migrating cells secrete proteins or even whole cell parts to mark their migration path and maybe to chemically attract following migrating cells. As pseudopodia such as filopodia play a probing role in 3D extracellular matrices, they have no function on 2D substrates, where the extracellular environment is more uniform (Fig. 4). The pseudopodial protrusion activity in 3D extracellular matrices is regulated by focal adhesion components. For example, the migration speeds of p130CAS knock-out cells and zyxin knock-out cells can be correlated with the number of protrusions such as filopodia generated per time in 3D extracellular matrices.[90] In more detail, the p130CAS knock-out cells move more slowly and zyxin knock-out cells more rapidly compared to their control wild-type cells in 3D extracellular matrices, whereas these knock-out cells exhibit the opposite motility phenotypes on 2D planar substrates (Fig. 4). The behavior of the vinculin knock-out cells has been shown to be similar to the migratory behavior of p130CAS knock-out cells depending on the dimensionality of the migration system.[50–52,103] Thus, the role of focal adhesion proteins in 2D motility systems cannot serve as a predictive model of their migratory role in the more physiologically relevant 3D migration systems. In particular, even the rate of filopodial protrusions formation does not seem to correlate with migration speed in 2D systems, whereas the rate of pseudopodial protrusions seems to be correlated with invasion speed in 3D migratory systems.[90] These results lead to the following questions: does protrusion dynamics play a role in 2D migratory systems? Does protrusion dynamics play solely a role in 3D migratory systems?

Recently, another invasion mode has been reported for cancer cells that possess a relatively soft cytoskeleton. This invasion mode seems to be different from the well-known mesenchymal and amoeboid migration modes. In more detail, these soft cells use a pulsating migration mode in which slow and random migration appears for a long time and is suddenly transformed to short-lived pulses of fast and directed migration which interrupt the slow migration mode.[74] Maybe this mode can be explained by the lobopodial invasion mode (Fig. 4), which has to be determined. These findings raise the question whether the EMT transition is still suitable for describing the migration of cancer cells in 3D matrices. Additionally, the soft cancer cells are surrounded by relatively stiff normal cells that migrate slow and a limited, little distance.[74] How this interaction between the soft cancer cells and the stiff normal cells may appear is suggested, but still under investigation. The fast migration periods of these special cancer cell migrations can be induced by myosin-II-dependent deformation of their soft nucleus evoked by the transient crowding of the neighboring normal cells with stiff nuclei.[74] Moreover, these neighboring stiffer normal cells can migrate due to cadherin-facilitated mismatch adhesions between normal cells and cancer cells (less cadherins), but their movement is limited by the residual α-catenin-mediated cell-cell adhesions such as homotypic E-cadherin adhesions between neighboring normal cells. These findings may explain the pulsating mode of cancer cell migration as these cancer cells cannot bind to normal cells and possess other mechanical properties.[74]

In addition, these mechanical properties of cancer cells do not solely affect their own functions such as cell motility, however, they also regulate the motility of normal cells such as fibroblasts, which are then no longer hindered in their migration through strong cell-cell adhesion bonds. However, while this precise regulatory mechanism is still under investigation, its recovery may shed light on the initial process of the cancer cell spreading from the primary tumor site. As the initial spreading of cancer cells from the primary tumor involves the crossing of the basement membrane of the tumor, this may be a similar process compared to the crossing of blood or lymph vessels where cancer cells have to overcome the basement membrane of the vessels first in order to transmigrate (intrastate) into the particular vessel by overcoming or even by breaking down the endothelial cell lining barrier of the vessels. This is discussed in more detail in the next section.

9. Transmigrating Invasive Cancer Cells Regulate the Biomechanical Properties of the Endothelial Cell Lining

The role of the endothelial cell lining of blood or lymph vessels on the regulation of cancer cell invasiveness into a 3D extracellular matrix is still elusive. The regulation of cancer cell transmigration is a complex scenario that is not yet fully characterized. In numerous previous studies, the endothelium has been reported to act as a passive barrier against the invasion of cancer cells.[104,105] In more detail, the endothelium has been found to decrease pronouncedly the invasion of cancer cells and hence, finally, cancer metastasis.[106] In contrast to these numerous reports, several recent reports propose a novel paradigm in which endothelial cells actively regulate the invasiveness of certain cancer cells by increasing their dissemination through vessels[107] or by enhancing the invasiveness of cancer cells into 3D matrices.[4] In particular, although several adhesion molecules have been identified to play a role in tumor-endothelial cell interactions and hence they even promote metastasis formation, however, the role of endothelial cell's mechanical properties during cancer cell transmigration and invasion are still elusive. It has been suggested that altered mechanical properties of endothelial cells may support one of its two main functions in cancer metastasis: they act either as a passive barrier or they serve as an active enhancer for cancer cell invasion. As a main biochemical pathway of the tumor-endothelial interaction, it has been reported that the involvement of cell adhesion receptors and integrins such as platelet endothelial cell adhesion molecule-1 (PECAM-1) and $\alpha v \beta 3$ integrins play a role, respectively.[108] As integrins are known to connect the extracellular matrix and the actomyosin cytoskeleton,[109–111] the linkage between the adhesion receptor and the actin cytoskeleton is facilitated through the mechano-coupling focal adhesion and cytoskeletal adaptor protein vinculin[51] and additionally determines the amount of cellular counter-forces that maintain the shape of the cells, their morphology and stiffness.[112] In particular, a broad biophysical approach to investigating the endothelial barrier break-down in the presence of co-cultured invasive cancer cells is still elusive. As microrheologic measurements such as magnetic tweezer rheology are well suited for the precise analysis of the endothelial cell's mechanical properties such as cellular stiffness during the co-culture with invasive or non invasive cancer cells compared to mono-cultured endothelial cells, endothelial stiffness is found to be influenced by co-cultured cancer cells.[69] In particular, highly-invasive breast cancer cells can influence the cellular mechanical properties of co-cultured human microvascular endothelial cells by reducing the stiffness of endothelial cells pronouncedly, whereas non-invasive cancer cells were not able to affect endothelial cell stiffness.[69] In addition, nanoscale particle tracking method diffusion measurements of actomyosin cytoskeletal-bound fibronectin-coated beads being markers for structural changes of the intercellular cytoskeletal scaffold can be used to measure the actomyosin-driven cytoskeletal remodeling dynamics. Thus, we find that cytoskeletal remodeling dynamics of endothelial cells are enhanced in co-culture with highly-invasive cancer cells, whereas they are even not altered in endothelial cells co-cultured with non-invasive cancer cells.[69] Finally, these findings indicate that highly-invasive breast cancer cells can alter actively the mechanical properties of co-cultured endothelial cells compared to monocultured endothelial cells, whereas non-invasive cancer cells were not able to alter the mechanical properties of endothelial cells. Thus, our results have provided for the first time an explanation for the breakdown of the endothelial barrier function of vessel wall monolayers and supported the special role of the neighboring endothelial cells surrounding primary tumors. Taken together, we have discussed how cancer cells can alter the mechanical properties in order to transmigrate through the endothelial cell lining of blood or lymph vessels. The next chapter raises the question or suggestion that endothelial cells may alter the mechanical properties of cancer cells are either reversible or non-reversible during their transmigration.

10. Do Endothelial Cells Alter the Mechanical Properties of Certain Invasive Cancer Cells?

Preliminary data of our group suggest that endothelial cells are indeed able to alter the mechanical properties of cancer cells. How long these alterations last have not yet been investigated. We hypothesize that these cancer cells, which transmigrated through the endothelial cell lining underwent massive morphological shape change followed by induction of signal transduction events after adhesion and transmigration through the endothelium. These may include the expression of genes that alter the mechanical properties of cancer cells. On the side of the endothelium, we have seen that these mechanical and biochemical alterations are there and seem to last longer than the duration of the whole transmigration process of cancer cells. Moreover, these alterations may also broadly affect the endothelial lining through a mechanically-driven signaling process across the endothelial monolayer.[113,114] On the side of the transmigrating cancer cell, the mechanical alterations of the endothelium may direct the cancer cell to the side of the transmigration process and may dictate the transmigration mode such as paracellular transmigration or transcellular transmigration. Whether this turns out to be true has to be further investigated in more detail.

11. Conclusions and Future Directions

The biomechanical interactions of cancer cells with their local microenvironment during the process of metastasis seems to be a key point in understanding the spreading of cancer cells from primary tumor sites and may also help to predict the overall survival rate of the patient more accurate. In particular, the physical and material properties of cancer cells regulate their migratory behavior and their transport through the human body after entering the blood or lymph vessels and hence, support or inhibit metastasis. Mechanical forces from the microenvironment may additionally regulate cancer cell motility (of epithelial origin) in the structurally complex extracellular matrix scaffold during invasion, intravasation and extravasation of cancer cells in and of the vascular system. Hence, insights into the role of physical and mechanical processes regulating metastasis can be a prerequisite for the development of new approaches for cancer diagnosis and treatment. Taken together, besides providing effective prognostic and diagnostic tools for therapies inhibiting metastasis, the knowledge of the role of biomechanics in cell motility may also inspire inverse strategies to promote wound healing in terms of connective tissue regeneration after injuries. The effects of key mechanical properties of the tumor microenvironment such as mechanical forces, stiffness, pore sizes and steric hindrances on cancer progression as well as the mechanical properties of stromal cells and endothelial cells on cancer cell invasion in general and after usage of therapeutic drugs have to be explored systematically. However, cutting-edge genetic or biochemical approaches need to be combined with novel and state-of-the-art biophysical measurements of cancer cell mechanics and the mechanical properties of the tissue microenvironment.

The effective combination of physics, molecular biology and biochemistry may provide the strength to reduce divergent effects of potential cancer drugs on cellular or organ responses in animal cancer disease models and cancer patients, and subsequently, may lead to more appropriate and efficient cancer treatments. The novel field of "physics of cancer" is currently rooted in biological physics and soft matter physics. The biophysics research is certainly more than simply serving as a sink for providing novel techniques for oncologists. Rather, it reveals novel aspects important for the understanding of cancer progression and helps to refine the functional pathways involved in cancer disease progression.

Acknowledgments

This work was supported by the Deutsche Krebshilfe (Grant No. 109432) and the DFG, which supported the 4th International Symposium on Physics of Cancer

(Grant No. MI1211/11-1). I thank Thomas M. L. Mierke for proof-reading and editing.

References

1. D. Hanahan and R. A. Weinberg, *Cell* **100**(1), 57–70 (2000).
2. D. Hanahan and R. A. Weinberg, *Cell* **144**, 646–674 (2011).
3. K. Bauer, C. Mierke and J. Behrens, *Int. J. Cancer* **121**, 1910–1918 (2007).
4. C. T. Mierke, D. P. Zitterbart, P. Kollmannsberger, C. Raupach, U. Schlotzer-Schrehardt, T. W. Goecke, J. Behrens and B. Fabry, *Biophys. J.* **94**, 2832–2846 (2008).
5. B. A. Teicher and S. P. Fricker, *Clin. Cancer Res.* **16**(11), 2927–2931 (2010).
6. J. Gong, D. Wang, L. Sun, E. Zborowska, J. K. Willson and M. G. Brattain, *Cell Growth Differ.* **8**, 83–90 (1997).
7. J. M. Ricono, M. Huang, L. A. Barnes, S. K. Lau, S. M. Weis, D. D. Schlaepfer, S. K. Hanks and D. A. Cheresh, *Cancer Res.* **69**, 1383–1391 (2009).
8. M. Z. Gilcrease, X. Zhou, X. Lu, W. A. Woodward, B. E. Hall and P. J. Morrissey, *J. Exp. Clin. Cancer Res.* **28**, 67 (2009).
9. K. Sawada, A. K. Mitra, A. R. Radjabi, V. Bhaskar, E. O. Kistner, M. Tretiakova, S. Jagadeeswaran, A. Montag, A. Becker, H. A. Kenny, M. E. Peter, V. Ramakrishnan, S. D. Yamada and E. Lengyel, *Cancer Res.* **68**, 2329–2339 (2008).
10. S. Runz, C. T. Mierke, S. Joumaa, J. Behrens, B. Fabry and P. Altevogt, *Biochem. Biophys. Res. Commun.* **365**, 35–41 (2008).
11. A. J. Garcia, F. Huber and D. Boettiger, *J. Biol. Chem.* **273**, 10988–10993 (1998).
12. Z. Gu, E. H. Noss, V. W. Hsu and M. B. Brenner, *J. Cell Biol.* **193**, 61–70 (2011).
13. S. S. Veiga, M. C. Q. B. Elias, W. Gremski, M. A. Porcionatto, R. da Silva, H. B. Nader and R. R. Brentani, *J. Biol. Chem.* **272**, 12529–12535 (1997).
14. J. Liu, X. He, Y. Qi, X. Tian, S. J. Monkley, D. R. Critchley, S. A. Corbett, S. F. Lowry, A. M. Graham and S. Li, *Mol. Cell. Biol.* **31**, 3366–3377 (2011).
15. P. T. Caswell, M. Chan, A. J. Lindsay, M. W. McCaffrey, D. Boettiger and J. C. Norman, *J. Cell Biol.* **183**, 143–155 (2008).
16. A. F. Chambers, A. C. Groom and I. C. MacDonald, *Nat. Rev. Cancer* **2**, 563–572 (2002).
17. P. S. Steeg, *Nat. Med.* **12**, 895–904 (2006).
18. J. Brábek, C. T. Mierke, D. Rösel, P. Veselý and B. Fabry, *Cell Commun. Signal.* **8**, 22 (2010).
19. C. T. Mierke, *Mol. Biosyst.* **8**, 1639–1649 (2012).
20. K. Wolf, I. Mazo, H. Leung, K. Engelke, U. H. von Andrian, E. I. Deryugina, A. Y. Strongin, E. B. Brocker and P. Friedl, *J. Cell. Biol.* **160**, 267–77 (2003).
21. C. T. Mierke, B. Frey, M. Fellner, M. Herrmann and B. Fabry, *J. Cell Sci.* **124**, 369–383 (2011).
22. C. T. Mierke, *New J. Phys.* **15**, 015003 (2013).
23. R. Kalluri and R. A. Weinberg, *J. Clin. Invest.* **119**, 1420–1428 (2009).
24. C. L. Chaffer and R. A. Weinberg, *Science* **331**, 1559–1564 (2011).
25. J. P. Thiery and J. P. Sleeman, *Nat. Rev. Mol. Cell Biol.* **7**, 131–142 (2006).
26. K. Polyak and R. A. Weinberg, *Nat. Rev. Cancer* **9**, 265–273 (2009).
27. D. J. Webb, K. Donais, L. A. Whitmore, S. M. Thomas, C. E. Turner, J. T. Parsons and A. F. Horwitz, *Nat. Cell Biol.* **6**, 154–161 (2004).
28. P. Friedl and E. B. Brocker, *Cell. Mol. Life Sci.* **57**, 41–64 (2000).

29. C. T. Mierke, D. Rosel, B. Fabry and J. Brabek, *Eur. J. Cell Biol.* **87**, 669–676 (2008).
30. C. T. Mierke, N. Bretz and P. Altevogt, *J. Biol. Chem.* **286**, 34858–34871 (2011).
31. M. H. Zaman, L. M. Trapani, A. L. Sieminski, D. Mackellar, H. Gong, R. D. Kamm, A. Wells, D. A. Lauffenburger and P. Matsudaira, *Proc. Natl. Acad. Sci. USA* **103**, 10889–10894 (2006).
32. K. Wolf, Y. I. Wu, Y. Liu, J. Geiger, E. Tam, C. Overall, M. S. Stack and P. Friedl, *Nat. Cell Biol.* **9**, 893–904 (2007).
33. P. Friedl and K. Wolf, *Cancer Metast. Rev.* **28**, 129–135 (2009).
34. X. Y. Li, I. Ota, I. Yana, F. Sabeh and S. J. Weiss, *Mol. Biol. Cell* **19**, 3221–3233 (2008).
35. Y. Itoh, N. Ito, H. Nagase and M. Seiki, *J. Biol. Chem.* **283**, 13053–13062 (2008).
36. S. Riedle, H. Kiefel, D. Gast, S. Bondong, S. Wolterink, P. Gutwein and P. Altevogt, *Biochem. J.* **420**(3), 391–402 (2009).
37. B. Singh, M. Schneider, P. Knyazev and A. Ullrich, *Int. J. Cancer* **124**, 531–539 (2009).
38. E. C. Bozkulak and G. Weinmaster, *Mol. Cell. Biol.* **29**, 5679–5695 (2009).
39. C. Brou, F. Logeat, N. Gupta, C. Bessia, O. LeBail, J. R. Doedens, A. Cumano, P. Roux, R. A. Black and A. Israel, *Mol. Cell* **5**, 207–216 (2000).
40. G. van Tetering, P. van Diest, I. Verlaan, E. van der Wall, R. Kopan and M. Vooijs, *J. Biol. Chem.* **284**, 31018–31027 (2009).
41. R. A. Black, C. T. Rauch, C. Kozlosky, J. J. Peschon, J. L. Slack and M. F. Wolfson, *Nature* **385**, 729–733 (1997).
42. D. Rösel, J. Brabek, O. Tolde, C. T. Mierke, D. P. Zitterbart, C. Raupach, K. Bicanova, P. Kollmannsberger, D.'Pankova, P. Vesely, P. Folkand and B. Fabry, *Mol. Cancer Res.* **6**, 410–420 (2008).
43. R. J. Bloom, J. P. George, A. Celedon, S. X. Sun and D. Wirtz, *Biophys. J.* **95**, 4077–4088 (2008).
44. T. M. Koch, S. Münster, N. Bonakdar, J. P. Butler and B. Fabry, *PLoS One* **7**(3), e33476 (2012).
45. N. Gjorevski and C. M. Nelson, *Biophys. J.* **103**(1), 152–162 (2012).
46. W. R. Legant, J. S. Miller, B. L. Blakely, D. M. Cohen, G. M. Genin and C. S. Chen, *Nat. Meth.* **7**, 969–971 (2010).
47. C. T. Mierke, *Cell Biochem. Biophys.* **66**, 599–622 (2013).
48. D. E. Discher, P. Janmey and Y. L. Wang, *Science* **310**, 1139–1143 (2005).
49. F. G. Giancotti, *Nat. Cell Biol.* **2**, E13–E14 (2000).
50. C. T. Mierke, *Cell Biochem. Biophys.* **53**, 115–126 (2009).
51. C. T. Mierke, P. Kollmannsberger, D. Paranhos-Zitterbart, J. Smith, B. Fabry and W. H. Goldmann, *Biophys. J.* **94**, 661–670 (2008).
52. C. T. Mierke, P. Kollmannsberger, D. P. Zitterbart, G. Diez, T. M. Koch, S. Marg, W. H. Ziegler, W. H. Goldmann and B. Fabry, *J. Biol. Chem.* **285**, 13121–13130 (2010).
53. S. P. Palecek, J. C. Loftus, M. H. Ginsberg, D. A. Lauffenburger and A. F. Horwitz, *Nature* **385**, 537–540 (1997).
54. J. C. Loftus and R. C. Liddington, *J. Clin. Invest.* **99**, 2302–2306 (1997).
55. A. R. Horwitz and J. T. Parsons, *Science* **286**, 1102–1103 (1999).
56. M. Rolli, E. Fransvea, J. Pilch, A. Saven and B. Felding-Habermann, *Proc. Nat. Acad. Sci. USA* **100**, 9482–9487 (2003).
57. R. S. Schmid, S. Shelton, A. Stanco, Y. Yokota, J. A. Kreidberg and E. S. Anton, *Development* **131**, 6023–6031 (2004).

58. S. C. Pawar, M. C. Demetriou, R. B. Nagle, G. T. Bowden and A. E. Cress, *Exp. Cell Res.* **313**, 1080–1089 (2007).
59. U. K. Rout, J. Wang, B. C. Paria and D. R. Armant, *Dev. Biol.* **268**, 135–151 (2004).
60. D. P. Ly, K. M. Zazzali and S. A. Corbett, *J. Biol. Chem.* **278**, 21878–21885 (2003).
61. C. T. Mierke, *Cell Biochem. Biophys.* **61**, 217–236 (2011).
62. K. Hotary, X. Y. Li, E. Allen, S. L. Stevens and S. J. Weiss, *Genes Dev.* **20**, 2673–2686 (2006).
63. K. B. Hotary, E. D. Allen, P. C. Brooks, N. S. Datta, M. W. Long and S. J. Weiss, *Cell* **114**, 33–45 (2003).
64. K. R. Levental, H. Yu, L. Kass, J. N. Lakins, M. Egeblad, J. T. Erler, S. F. Fong, K. Csiszar, A. Giaccia, W. Weninger, M. Yamauchi, D. L. Gasser and V. M. Weaver, *Cell* **139**, 891–906 (2009).
65. O. De Wever, P. Demetter, M. Mareel and M. Bracke, *Int. J. Cancer* **123**, 2229–2238 (2008).
66. P. P. Provenzano, D. R. Inman, K. W. Eliceiri and P. J. Keely, *Oncogene* **28**, 4326–4343 (2009).
67. J. K. Mouw, Y. Yui, L. Damiano, R. O. Bainer, J. N. Lakins, I. Acerbi, G. Ou, A. C. Wijekoon, K. R. Levental, P. M. Gilbert, E. S. Hwang, Y. Y. Chen and V. M. Weaver, *Nat. Med.* **20**(4), 360–367 (2014).
68. A. J. Engler, S. Sen, H. L. Sweeney and D. E. Discher, *Cell* **126**, 677–689 (2005).
69. C. T. Mierke, *J. Biol. Chem.* **286**, 40025–40037 (2011).
70. C. T. Mierke, *Phys. Biol.* **10**(6), 065005 (2013).
71. S. Kumar and V. Weaver, *Cancer Metast. Rev.* **28**, 113–127 (2009).
72. M. J. Paszek, N. Zahir, K. R. Johnson, J. N. Lakins, G. I. Rozenberg, A. Gefen, C. A. Reinhart-King, S. S. Margulies, M. Dembo, D. Boettiger, D. A. Hammer and V. W. Weaver, *Cancer Cell* **8**(3), 241–254 (2005).
73. A. Parekh and A. M. Weaver, *Cell Adh. Migr.* **3**(3), 288–292 (2009).
74. M. H. Lee, P. H. Wu, J. R. Staunton, R. Ros, G. D. Longmore and D. Wirtz, *Biophys. J.* **102**(12), 2731–2741 (2012).
75. C. Storm, J. J. Pastore, F. C. MacKintosh, T. C. Lubensky and P. A. Janmey, *Nature* **435**, 191–194 (2005).
76. K. Franke, J. Sapudom, L. Kalbitzer, U. Anderegg and T. Pompe, *Acta Biomater.* **10**(6), 2693–2702 (2014).
77. M. R. Ng and J. S. Brugge, *Cancer Cell* **16**(6), 455–457 (2009).
78. K. M. Branch, D. Hoshino and A. M. Weaver, *Biol. Open* **1**(8), 711–722 (2012).
79. B. Geiger, A. Bershadsky, R. Pankov and K. M. Yamada, *Nat. Rev. Mol. Cell Biol.* **2**, 793–805 (2001).
80. A. K. Harris, D. Stopak and P. Wild, *Nature* **290**, 249–251 (1981).
81. J. Guck, S. Schinkinger, B. Lincoln, F. Wottawah, S. Ebert, M. Romeyke, D. Lenz, H. M. Erickson, R. Ananthakrishnan, D. Mitchell, J. Käs, S. Ulvick and C. Bilby, *Biophys. J.* **88**, 3689–3698 (2005).
82. J. T. Parsons, A. R. Horwitz and M. A. Schwartz, *Nat. Rev. Mol. Cell Biol.* **11**, 633–643 (2010).
83. A. J. Ridley, M. A. Schwartz, K. Burridge, R. A. Firtel, M. H. Ginsberg, G. Borisy, J. T. Parsons and A. R. Horwitz, *Science* **302**, 1704–1709 (2003).
84. T. D. Pollard and G. G. Borisy, *Cell* **112**, 453–465 (2003).
85. D. A. Lauffenburger and A. F. Horwitz, *Cell* **84**, 359–369 (1996).
86. L. D. Wood, D. W. Parsons, S. Jones, J. Lin, T. Sjöblom and R. J. Leary, *Science* **318**, 1108–1113 (2007).
87. F. Sabeh, R. Shimizu-Hirota and S. J. Weiss, *J. Cell Biol.* **185**, 11–19 (2009).

88. N. E. Sounni, L. Devy, A. Hajitou, F. Frankenne, C. Munaut, C. Gilles, C. Deroanne, E. W. Thompson, J. M. Foidart and A. Noel, *FASEB J.* **16**, 555–564 (2002).
89. A. S. Adhikari, J. Chai and A. R. Dunn, *J. Am. Chem. Soc.* **133**, 1686–1689 (2011).
90. S. I. Fraley, Y. Feng, R. Krishnamurthy, D. H. Kim, A. Celedon, G. D. Longmore and D. Wirtz, *Nat. Cell Biol.* **12**, 598–604 (2010).
91. M. A. Wozniak, R. Desai, P. A. Solski, C. J. Der and P. J. Keely, *J. Cell Biol.* **163**, 583–595 (2003).
92. D. Yamazaki, S. Kurisu and T. Takenawa, *Oncogene* **28**, 1570–1583 (2009).
93. A. D. Doyle, F. W. Wang, K. Matsumoto and K. M. Yamada, *J. Cell Biol.* **184**, 481–490 (2009).
94. D. A. Murphy and S. A. Courtneidge, *Nat. Rev. Mol. Cell Biol.* **12**(7), 413–426 (2011).
95. M. L. Smith, D. Gourdon, W. C. Little, K. E. Kubow, R. A. Eguiluz, S. Luna-Morris and V. Vogel, *PLoS Biol.* **5**, e268 (2007).
96. S. X. Sun, S. Walcott and C. W. Wolgemuth, *Curr. Biol.* **20**, R649–R654 (2010).
97. W. T. Shih and S. Yamada, *Biophys. J.* **98**, L29–L31 (2010).
98. F. Rehfeldt, A. E. X. Brown, M. Raab, S. Cai, A. L. Zajac, A. Zemelc and D. E. Discher, *Integr. Biol.* **4**, 422–430 (2012).
99. K. A. Beningo, M. Dembo, I. Kaverina, J. V. Small and Y. L. Wang, *J. Cell Biol.* **153**, 881–888 (2001).
100. T. Lämmermann, B. L. Bader, S. J. Monkley, T. Worbs, R. Wedlich-Söldner, K. Hirsch, M. Keller, R. Förster, D. R. Critchley, R. Fässler and M. Sixt, *Nature* **453**, 51–55 (2008).
101. S. Even-Ram and K. M. Yamada, *Curr. Opin. Cell Biol.* **17**, 524–532 (2005).
102. A. D. Doyle, F. W. Wang, K. Matsumoto and K. M. Yamada, *J. Cell Biol.* **184**, 481–490 (2009).
103. G. S. Goldberg, D. B. Alexander, P. Pellicena, Z.-Y. Zhang, H. Tsuda and W. T. Miller, *J. Biol. Chem.* **278**, 46533–46540 (2003).
104. A. B. Al-Mehdi, K. Tozawa, A. B. Fisher, L. Shientag, A. Lee and R. J. Muschel, *Nat. Med.* **6**, 100–102 (2000).
105. A. Zijlstra, J. Lewis, B. Degryse, H. Stuhlmann and J. P. Quigley, *Cancer Cell* **13**, 221–234 (2008).
106. G. L. Van Sluis, T. M. Niers, C. T. Esmon, W. Tigchelaar, D. J. Richel, H. R. Buller, C. J. Van Noorden and C. A. Spek, *Blood* **114**, 1968–1973 (2009).
107. D. Kedrin, B. Gligorijevic, J. Wyckoff, V. V. Verkhusha, J. Condeelis, J. E. Segall and J. van Rheenen, *Nat. Meth.* **5**, 1019–1021 (2008).
108. E. B. Voura, N. Chen and C. H. Siu, *Clin. Exp. Metastasis* **18**, 527–532 (2000).
109. N. T. Neff, C. Lowrey, C. Decker, A. Tovar, C. Damsky, C. Buck and A. F. Horwitz, *J. Cell Biol.* **95**, 654–666 (1982).
110. C. H. Damsky, K. A. Knudsen, D. Bradley, C. A. Buck and A. F. Horwitz, *J. Cell Biol.* **100**, 1528–1539 (1985).
111. D. Riveline, E. Zamir, N. Q. Balaban, U. S. Schwarz, T. Ishizaki, S. Narumiya, Z. Kam, B. Geiger and A. D. Bershadsky, *J. Cell Biol.* **153**, 1175–1186 (2001).
112. A. D. Rape, W. H. Guo and Y. L. Wang, *Biomaterials* **32**, 2043–2051 (2011).
113. C. Raupach, D. Paranhos-Zitterbart, C. Mierke, C. Metzner, A. F. Müller and B. Fabry, *Phys. Rev. E* **76**, 011918 (2007).
114. C. Metzner, C. Raupach, C. T. Mierke and B. Fabry, *J. Phys.: Conden. Mat.* **22**, 194105 (2010).

Chapter 3

Advances in Computational Radiation Biophysics for Cancer Therapy: Simulating Nano-Scale Damage by Low-Energy Electrons

Zdenka Kuncic

School of Physics, The University of Sydney
Sydney, NSW 2006, Australia
zdenka.kuncic@sydney.edu.au

Computational radiation biophysics is a rapidly growing area that is contributing, alongside new hardware technologies, to ongoing developments in cancer imaging and therapy. Recent advances in theoretical and computational modeling have enabled the simulation of discrete, event-by-event interactions of very low energy ($\ll 100\,\mathrm{eV}$) electrons with water in its liquid thermodynamic phase. This represents a significant advance in our ability to investigate the initial stages of radiation induced biological damage at the molecular level. Such studies are important for the development of novel cancer treatment strategies, an example of which is given by microbeam radiation therapy (MRT). Here, new results are shown demonstrating that when excitations and ionizations are resolved down to nano-scales, their distribution extends well outside the primary microbeam path, into regions that are not directly irradiated. This suggests that radiation dose alone is insufficient to fully quantify biological damage. These results also suggest that the radiation cross-fire may be an important clue to understanding the different observed responses of healthy cells and tumor cells to MRT.

1. Introduction

The first Nobel prize in Physics was awarded to Willhelm Röntgen in 1901 for the discovery of x-ray radiation. Almost immediately, x-rays were being used to image parts of the human body and within just a few years of their discovery, x-rays were being used to treat cancers and other malignancies. Today, the use of radiation is pervasive throughout medicine, both in various medical imaging procedures as well as radiation cancer treatment, and innovations in medical radiation technologies continue to drive more sophisticated approaches to diagnosing and treating cancer.

Radiation cancer therapy, which is used to treat up to 50% of all cases worldwide,[1] is premised on eradication of tumor cells by irreparable DNA damage. Strand breaks and other lesions result from radiation interactions (ionizations and excitations) that occur on or near genes whose primary function is to encode proteins that regulate the cell cycle. At sufficiently high doses, radiation-induced DNA damage leads to a well-known deterministic response: cell death.[2,3]

Computational simulations offer a powerful (and cost-effective) tool for quantitatively investigating radiation interactions with biological tissue. The Monte Carlo numerical technique has been extensively used for simulating the stochastic interactions of radiation particles (photons, electrons, protons and heavier ions) with relevant biological molecules (e.g. DNA, H_2O).[4] As radiation damage can be largely attributed to the secondary low-energy electrons produced in copious quantities from primary radiation interaction events, Monte Carlo radiation transport simulations rely crucially on accurate cross sections for inelastic scatter of electrons at energies comparable to the energy levels of biomolecules (typically a few tens of eV).[5] A major challenge has been improving the accuracy of these low-energy electron cross sections for liquid water, as a surrogate for the soft condensed nature of biological tissue.[6,7]

This article reviews theoretical developments in modeling inelastic scatter of low-energy ($\ll 100\,\mathrm{eV}$) electrons in liquid water. New simulation results are also presented demonstrating how the implementation of these interaction cross sections into Monte Carlo radiation transport codes has revealed new insights into an experimental cancer treatment technique, microbeam radiation therapy.

2. Theoretical Background and Computational Approach

Copius secondary electrons produced as a result of primary interaction events are chiefly responsible for the ensuing radiation damage caused by excitations and ionizations.[8] Better theoretical models are, however, needed for the interaction of low energy ($\ll 100\,\mathrm{eV}$) electrons in soft condensed matter. This section describes the relevant theoretical background in this area and some recent developments in computational models.

2.1. *Low-energy electron inelastic scatter in liquid water*

Scattering theory provides the theoretical framework for calculating the probability of inelastic scatter by electrons off an atomic or molecular target.[9,10] A major challenge in an *ab initio* analytical approach, however, is a sufficiently accurate description of the molecular wavefunction of the target, which is a multi-electron system (e.g. liquid water).[5,6] Experimental scatter experiments effectively probe the structure of the target and thus provide a means to directly infer the scatter probability.[11] It is then convenient to express the differential scatter cross-section in terms of the dynamic structure factor,[12,13] which is effectively the quantity that is measured in scatter experiments.[14,15] The dynamic structure factor is also directly related to the dielectric response function, which offers a somewhat more tractable analytic approach and has thus been used to complete cross-section data tables for very low energy electron inelastic scatter in liquid water.[16]

2.1.1. *Dynamic structure factor*

In the first Born approximation, the differential cross section for inelastic electron scattering off an atomic or molecular target containing bound charges with respect to solid scattering angle Ω' and energy of scattered electrons E' is[9]:

$$\frac{d^2\sigma}{d\Omega'dE'} = \frac{mp'}{4\pi^2\hbar^4}|\langle\Psi_f|V|\Psi_i\rangle|^2\delta(E' + E_n - E_0) \tag{1}$$

where i and f refer to the initial and final states, respectively, of the system (electron plus target), V is the interaction potential, p' is the momentum of the scattered electron, and E_0 and E_n are the corresponding energies of the atom/molecule. Assuming plane waves for the electrons, (1) can be expressed as

$$\frac{d^2\sigma}{d\Omega'dE'} = \frac{m^2}{4\pi^2\hbar^4}\frac{p'}{p}\left|\int\int V(\mathbf{r})\exp(-i\mathbf{q}\cdot\mathbf{r})\psi_n^*\psi_0 d^3\mathbf{r_j}d^3\mathbf{r}\right|^2\delta(E' + E_n - E_0) \tag{2}$$

where $\mathbf{q} = \mathbf{p} - \mathbf{p}'$ is the momentum transfer and $d^3\mathbf{r_j}$ is the differential volume of the target containing j bound electrons. Noting that the interaction potential is coulombic, with $V(\mathbf{r}) = \sum_j e^2/|\mathbf{r} - \mathbf{r}_j|$, yields

$$\frac{d^2\sigma}{d\Omega'dE'} = \frac{m^2}{4\pi^2\hbar^5}\frac{p'}{p}[V(\mathbf{q})]^2\left|\int\psi_n^*\sum_j\exp(-i\mathbf{q}\cdot\mathbf{r}_j)\psi_0 d^3\mathbf{r_j}\right|^2\delta\left(\frac{E'}{\hbar} + \frac{E_n - E_0}{\hbar}\right) \tag{3}$$

where $V(\mathbf{q})$ is the Fourier transform of $V(\mathbf{r})$. This is now in the same form given in Ref. 12. Expressing the delta function in terms of its Fourier representation, this can be further simplified to

$$\frac{d^2\sigma}{d\Omega'dE'} = \frac{m^2}{4\pi^2\hbar^5}\frac{p'}{p}[V(\mathbf{q})]^2 S(\omega, \mathbf{q}) \tag{4}$$

where $S(\omega, \mathbf{q})$ is the dynamic structure factor, defined as

$$S(\omega, \mathbf{q}) = \frac{1}{(2\pi)^4}\int\int_{-\infty}^{+\infty} dt\, d^3\mathbf{r}\exp[i(\mathbf{q}\cdot\mathbf{r} - \omega t)]$$

$$\times\sum_{j,l}^N\int d\mathbf{q}\exp(-i\mathbf{q}\cdot\mathbf{r})\langle\exp[-i\mathbf{q}\cdot\mathbf{r}_l(0)]\exp[i\mathbf{q}\cdot\mathbf{r}_j(t)]\rangle. \tag{5}$$

In this form, it is evident that $S(\omega, \mathbf{q})$ is equivalent to the Fourier transform of the spatial correlation of charges in the target.[17] This can be re-expressed more simply as

$$S(\omega, \mathbf{q}) = \frac{1}{2\pi}\sum_{j,l}^N\int_{-\infty}^{+\infty} dt\exp(-i\omega t)\langle\exp[-i\mathbf{q}\cdot\mathbf{r}_l(0)]\exp[i\mathbf{q}\cdot\mathbf{r}_j(t)]\rangle. \tag{6}$$

2.1.2. Dielectric response approach

The structure factor is a valuable quantity because it is in principle directly measurable through scatter experiments.[14,15] For biological (i.e. soft condensed matter) targets, electron/positron scatter experiments present a difficult challenge because the vacuum conditions required to mitigate beam scatter are incompatible with maintaining a liquid phase of the sample. X-ray or neutron scatter experiments are thus more favourable for probing the electronic structure of molecular targets in the condensed phase (e.g. liquid water). Theoretical models for low-energy ($\ll 100\,\text{eV}$) inelastic electron scatter cross-sections for biological molecules in the condensed phase are often derived in the framework of linear response theory and first-order perturbation theory.[9,10] In that framework, the relevant quantity is the dielectric response function $\epsilon(\omega, \mathbf{q})$. Quantum mechanically, $\epsilon(\omega, \mathbf{q})$ is represented in terms of matrix elements between exact eigenstates of the many-body system. $\epsilon(\omega, \mathbf{q})$ can also be derived from classical electrodynamics using the concept of Drude oscillators.[17]

The fluctuation-dissipation theorem[18] establishes a relation between the structure factor and the dielectric response: the correlation in electron density fluctuations is directly proportional to the dissipation of energy due to a perturbing force. Reference 17 gives the expression for $1/\epsilon(\omega, \mathbf{q})$. The relevant quantity of interest is the imaginary part, which determines energy transfer to the target:

$$\text{Im}\left[\frac{1}{\epsilon(\omega, \mathbf{q})}\right] = \frac{4\pi^2 e^2 n}{\hbar q^2} \frac{1}{2\pi} \sum_{j,l}^{N} \int_{-\infty}^{+\infty} dt\, (\exp(i\omega t)$$

$$- \exp(-i\omega t))\, \langle \exp[-i\mathbf{q} \cdot \mathbf{r}_l(0)] \exp[i\mathbf{q} \cdot \mathbf{r}_j(t)] \rangle \qquad (7)$$

where n is the number density of electrons in the target. From a comparison with (5), we can re-write this as

$$S(\omega, \mathbf{q}) = \frac{\hbar q^2}{4\pi^2 e^2 n} \text{Im}\left[\frac{-1}{\epsilon(\omega, \mathbf{q})}\right]. \qquad (8)$$

Thus, the differential cross-section for inelastic electron scatter (4) can be expressed in terms of the dielectric response function as follows:

$$\frac{d^2\sigma}{d\Omega' dE'} = \frac{m^2 q^2}{(4\pi^2)^2 \hbar^4 e^2 n} \frac{p'}{p} [V(\mathbf{q})]^2 \, \text{Im}\left[\frac{-1}{\epsilon(\omega, \mathbf{q})}\right]. \qquad (9)$$

Substituting $V(\mathbf{q}) = 4\pi e^2/q^2$ for a Coulomb interaction potential yields

$$\frac{d^2\sigma}{d\Omega' dE'} = \frac{m^2}{\pi^2 \hbar^4 q^2 n} \frac{p'}{p} \text{Im}\left[\frac{-1}{\epsilon(\omega, \mathbf{q})}\right]. \qquad (10)$$

It is common for this to be given in terms of the differential inverse mean free path (DIMP), $\Lambda = n\sigma$ and for the second-order differential to be represented in terms of dq rather than $d\Omega$. To derive this, we use relations $q = |\mathbf{p} - \mathbf{p}'|$ and $E = \hbar^2 q^2/2m$, so that

$$q = \frac{\sqrt{2m}}{\hbar}[2T - \hbar\omega - 2\sqrt{T(T - \hbar\omega)}\cos\theta]^{1/2}$$

where $\hbar\omega = T - E'$ is the energy transfer from incident electron with initial energy T and $\theta = \cos^{-1}(\mathbf{p}' \cdot \mathbf{p})/p'p$ is the scatter angle. Using $d\Omega' = d\cos\theta d\phi$ and calculating $dq/d\theta$ explicitly gives the following relation

$$d\Omega' = \frac{-\hbar}{\sqrt{2m}} \frac{[2T - \hbar\omega - 2\sqrt{T(T - \hbar\omega)}\cos\theta]^{1/2}}{\sqrt{T(T - \hbar\omega)}} dq d\phi.$$

After some algebra, we can rewrite (10) as

$$\frac{d^2\Lambda}{dqdE'} = \frac{1}{\pi a_0 Tq} \text{Im}\left[\frac{-1}{\epsilon(\omega, \mathbf{q})}\right] \tag{11}$$

where $a_0 = \hbar^2/me^2$ is the Bohr radius.

This equation for the DIMP provides a theoretical basis for the models for low energy ($\lesssim 100\,\text{eV}$) electron inelastic scatter in water that are used in modern Monte Carlo radiation transport codes.[16,19,20] For liquid water, which best represents the soft condensed matter phase of biological tissue, the dielectric response function should take into account the excitation and ionization energy levels in that thermodynamic phase. Experimental data for scattering off a liquid water target are scarce and the available data that is used to validate Drude oscillator models for $\epsilon(\omega, \mathbf{q})$ is in the optical limit, $q \to 0$.[15,21] Although some liquid water structure function data has been published for x-ray scatter with finite momentum transfer,[14] the approach used to date has been to extrapolate the theoretical model for $\epsilon(\omega, 0)$ into the $q > 0$ domain (the so-called Bethe ridge — cf. Refs. 16, 20–22).

2.2. *Monte Carlo radiation transport modeling*

Several Monte Carlo radiation transport codes have been developed for simulating radiation interactions with materials of biological interest. The key differences between the different codes can be largely attributed to the different application endpoints, which have driven the development of certain capabilities over others (e.g. development of neutron and photon transport over proton and heavy ion transport).

The Monte Carlo N-Particle (MCNP) transport code[a] has the longest history of usage, having been developed by Los Alamos National Laboratory more than 50 years ago. Although it is now a general purpose Monte Carlo code, its development has focused mainly towards applications to nuclear processes. The electron gamma shower (EGS) Monte Carlo software tool was originally developed in the 1980's by

---

[a] mcnp.lanl.gov

the National Research Council (NRC) Canada and the Stanford Linear Accelerator Center (SLAC).[23] EGSnrc simulations electron-photon transport over particle energies $1\,\mathrm{keV} - 10\,\mathrm{GeV}$ relevant specifically to medical radiation applications.[24] One particularly important application is to the simulation of medical linear accelerators used for cancer radiotherapy. PARTRAC is a sophisticated Monte Carlo radiation transport toolkit specifically developed to simulate electron and ion track structure through biological material and the radiation chemistry relevant to ensuing biological damage.[25] Many other more specialized track structure codes have also been developed for microdosimetry and radiation chemistry applications (see Ref. 26 for a review).

This article describes radiation transport simulations using Geant4. Geant4 is an open-source software toolkit based on Monte Carlo code that was originally developed by CERN for high-energy particle physics applications. It has since been extended considerably to simulate lower-energy ($\ll 1\,\mathrm{GeV}$) particle interactions relevant for medical and biological radiation physics applications.[27] An important recent development is the extension of electron inelastic scatter cross-sections in liquid water down to a theoretical limit of zero eV, based on theoretical models for the dielectric response function $\epsilon(\omega, \mathbf{q})$ (cf. Sec. 2.1.2) and available experimental data. These same electron inelastic cross-section models are also implemented in PARTRAC.[20] In Geant4, the liquid-water cross-section models are available in the module known as Geant4-DNA.[28]

2.2.1. *Geant4-DNA*

Geant4-DNA represents an ongoing international collaboration[b] aimed at improving models for nano-scale radiation interactions with DNA, H_2O and other key biomolecules. It is currently able to simulate a range of low-energy ($\ll 100\,\mathrm{eV}$) electron interactions (elastic scatter, electronic excitation, vibrational excitation, ionisation, attachment) and also some low-energy proton and alpha particle interactions.[29] All low-energy interactions are currently in liquid water only. Efforts are underway to extend this capability to other biomolecules and to simulate the radiation chemistry processes involving free radicals that are responsible for biological damage.[30] Because low-energy electrons are largely responsible for generating free radicals and thus initiating damage, Geant4-DNA explicitly simulates discrete electron interactions. Numerically, this is achieved by extending the electronic stopping power tables down to very low energies, extrapolating down to zero eV. Figure 1 shows the track lengths and penetration depth (in nanometers) used by Geant4-DNA, calculated from stopping powers based on the theoretical cross-section calculations and experimental data discussed in the previous section. See Ref. 31 for further details.

---

[b]geant4-dna.org

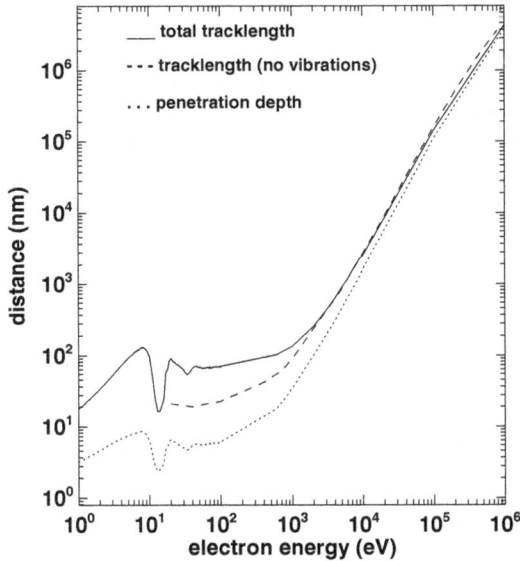

Fig. 1.   Electron track length and penetration depth in liquid water as a function of initial electron energy calculated using Geant4-DNA. (Data provided by Z. Francis, Geant4-DNA collaboration.)

3. A Case Study: Microbeam Radiation Therapy

Cancer radiation therapy using synchrotron x-ray microbeams has been proposed as a next-generation treatment strategy. Although still an experimental technique, results to date are promising, showing a higher therapeutic index (ratio of curative effects to adverse normal tissue effects) than conventional radiotherapy using a medical linear accelerator beam (see Ref. 32 and references therein). Microbeam radiation therapy (MRT) makes use of high-intensity, highly focused (typically ~ 20–$50\,\mu$m) synchrotron beamlets arranged in a planar array configuration. MRT small animal studies have demonstrated that the sharp spatial modulation in delivered radiation affords a high efficacy of tumour cell kill with relatively low adverse effects in surrounding normal tissue. Although the underlying mechanisms for the effectiveness of MRT remain to be convincingly elucidated, a contributing factor may be that a high resilience of normal cells to the radiation enables these cells to be repaired and regenerated as a result of intercellular signalling involving neighboring normal cells not directly exposed to the radiation.[33]

3.1. *Simulating nano-scale electron interactions with Geant4-DNA*

To gain more physical insight into MRT, simulations were carried out using Geant4-DNA. Of particular interest was the total number of ionizations and excitations as a more pertinent indicator of biological damage than radiation dose. Figure 2 shows the geometry setup in the Geant4 simulation. A $16 \times 16 \times 16\,\mathrm{cm}^3$ cubic water phantom was irradiated with monoenergetic 50 keV x-rays with a spatial distribution of

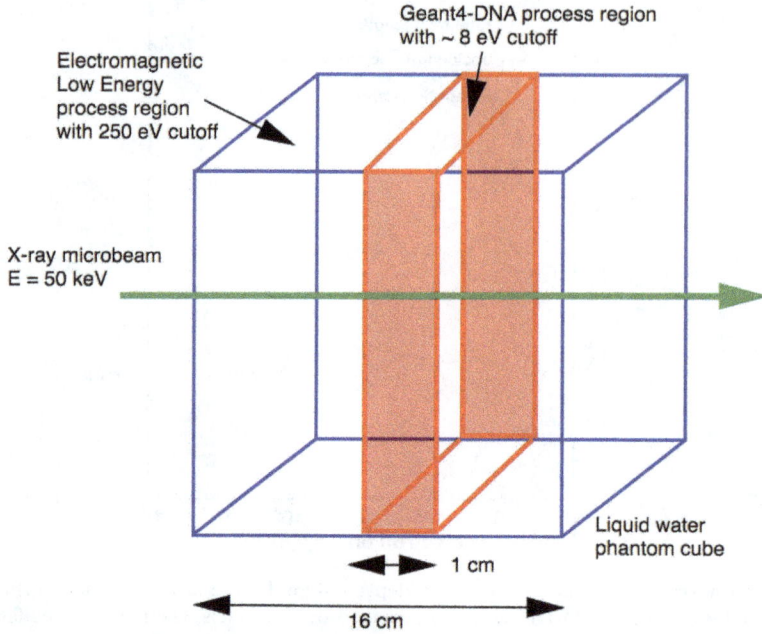

Fig. 2.   Geometry of the Geant4 x-ray microbeam simulation. The red region shows a slice in which Geant4-DNA was used to explicitly simulate event-by-event electron interactions. (Figure provided by A. McNamara, University of Sydney.)

$5 \times 20\,\mu$m, consistent with the typical size of a microbeam array slit. Interaction processes modelled in the simulation included: the photoelectric effect, Compton scattering, Rayleigh scattering, and electron scattering (elastic and inelastic). A 1 cm thick cross-sectional slice was selected at the entrance and exit sides of the irradiated phantom and the Geant4-DNA processes were switched on in these two regions to simulate discrete, event-by-event electron interactions, including ionizations as well as electronic and vibrational excitations. In the remainder of the water phantom, the Geant4 low-energy electromagnetic models were used, which are valid down to a cutoff energy of 250 eV.

Figure 3 shows results of the Geant4 x-ray microbeam simulations. The incidence of all interaction processes simulated is shown, together with a 3D visualization of electron track structure in the water phantom. The electron tracks are very short (typically no more than a few microns in length) and thus appear as points on the phantom scale. A higher density of electron tracks is evident near each end of the phantom where the microbeam enters and exits, corresponding to the 1 cm thickness slice in which Geant4-DNA was used to resolve event-by-event electron interactions down to nano-scales. The corresponding electron tracks in this subregion are distributed in distinctive localized clumps or clusters situated away from the primary x-ray microbeam path. This dispersed clustering can be attributed to Compton scattering of primary x-rays out of the beam and the subsequent production of

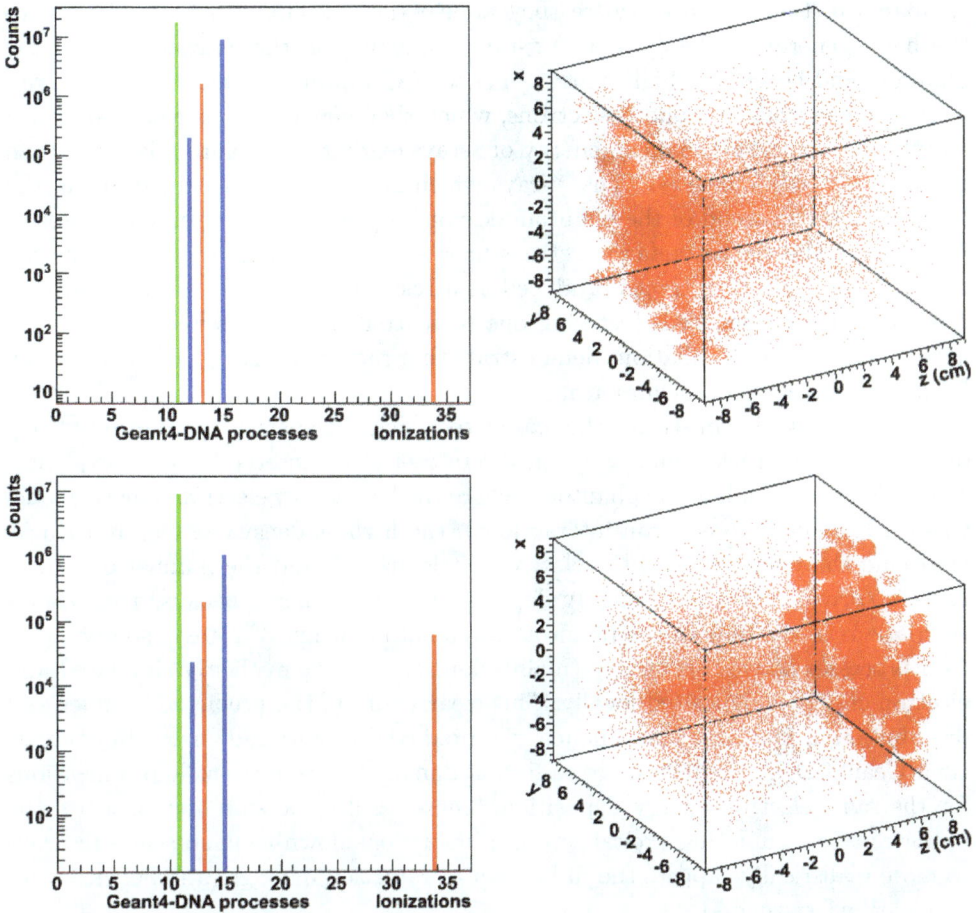

Fig. 3. (Color online) Geant4 simulations of x-ray microbeam interactions in a water phantom, with 50 keV x-rays entering at $z = -8$ cm (cf. Fig. 2). The right column shows secondary electron tracks (red) calculated using two different models in two distinct regions in the phantom: the Geant4-DNA event-by-event model, used in a 1 cm layer at the entrance (top right) and exit (bottom right) region; and the Geant4 low-energy electromagnetic model (with a 250 eV cutoff), used in the remaining parts of the phantom. The left column shows frequency histograms of the electron interaction processes throughout the whole phantom, consisting of Geant4-DNA processes (nos. 11–15) used in the entrance/exit layers and ionizations (process no. 34) calculated by the low-energy electromagnetic model. The Geant4-DNA electron processes shown are: elastic scatter (green), excitation (red) and ionization (blue). A total of 10^6 incident x-rays were used in each simulation. (Figure provided by A. McNamara, University of Sydney.)

successive generations of secondary electrons from additional Compton scatter and photoelectron processes, as well as inelastic electron scatter. Most of the electron track clusters are located at a distance $\simeq 5$ cm from the primary beam, which is consistent with the mean free path for Compton scatter of 50 keV x-rays in water (the probability of photoelectric absorption is approximately 7 times smaller at this energy). The electron track clusters appear more diffuse in the entrance slice compared to the exit slice, where they appear denser. This can be attributed to the higher energy of the (primary) x-rays impinging on the entrance side of the phantom and hence, the higher mean energy of Compton scattered x-rays, Compton recoil electrons and photo-electrons, which then generate longer electron track lengths. By contrast, the mean energy of x-rays exiting the phantom is lower than the incident 50 keV energy. Thus, x-rays which are Compton scattered out of the beam near the exit end of the phantom do not propagate as far between successive interactions and the secondary electrons have a shorter average range. The spatial distribution of electron tracks produced as a result has a smaller variance, with a greater degree of localised clustering due to scatter of electrons with increasingly lower energy. These simulations demonstrate that radiation damage can occur away from the primary microbeam path.

Figure 3 also demonstrates the extent to which the spatial distribution of secondary electron tracks changes when discrete electron interactions are explicitly simulated. In the bulk of the phantom (where the low-energy electromagnetic model was used to simulate electron interactions), the highest density of electron tracks is around the microbeam, with relatively little spread, and the number of ionizations is correlated with the energy deposition (dose). This is because the Geant4 simulations in the bulk of the phantom use a cutoff energy of 250 eV, so only electron processes above this energy (mainly ionizations) are explicitly simulated and electron energy is deposited locally. This clearly affects the predicted number and distribution of ionizations and hence, the predicted radiation damage. In comparison, Geant4-DNA predicts more radiation damage by excitations and ionizations for the same electron energy deposition. These results therefore demonstrate the importance of simulating excitations and ionizations down to nano-scales in order to more realistically capture the full extent of radiation damage (and hence, radiobiological effectiveness).

What are the implications of these results for microbeam radiation therapy (MRT)? The results indicate that when the excitations and ionizations responsible for radiation damage are resolved down to the nano-scale, the peak and valley dose pattern on the micron-scale does not accurately reflect the underlying physical action of radiation. The observed biological response to MRT suggests that the spread of low-level excitations and ionizations into the unirradiated valley regions may be essential for priming the cells therein for damage response. Ionizations in particular produce reactive free radicals that trigger a cascade of damage-sensing biochemical signalling pathways. In the valley regions, healthy cells primed by the low-level damage are able to coordinate an effective repair response and intercellular signalling to neighbouring healthy cells in the peak dose regions results in successful repair and regeneration. In contrast, tumor cells which are damaged by ionization cross-fire in the valley regions are unable to coordinate an effective repair response and thus succumb to the same fate as their neighbouring tumor cells in the peak dose regions. In the absence of any radiation exposure in the valley regions, tumor cells therein may persist and continue proliferating.

These results also have implications for the bystander effect, which describes the phenomenon whereby unirradiated cells exhibit damage responses similar to neighbouring irradiated cells. As this study has shown, cells that are not directly irradiated are still at risk of damage by radiation cross-fire. A zero dose can be misleading, even when measured on micron-scales, because it does not accurately reflect the full extent of radiation damage caused by excitations and ionizations on nano-scales.

4. Conclusion

Computational radiation biophysics is proving to be a powerful method for testing our understanding of the basic tenets underpinning cancer radiotherapy and radiobiology. Ongoing theoretical developments in improving the interaction cross-sections of the very low energy electrons responsible for biological damage are now beginning to take into account the condensed phase and molecular nature of biological targets, in particular, liquid water. These theoretical models have been successfully implemented into a new generation of Monte Carlo radiation transport codes, which are enabling more realistic simulations of radiation interactions with biological matter for medical applications. For cancer therapy, in particular, this new simulation capability has already revealed new insights into the experimental technique of microbeam radiation therapy. Results shown here have revealed that when electron interactions are resolved down to nano-scales, the extent of radiation damage is significantly greater than what is inferred from deposited dose. The distribution of ionizations extends well beyond the primary microbeam, spreading into surrounding regions that are not directly irradiated. Thus, dose alone is inadequate for describing radiobiological effectiveness. Continued improvements in the efficacy of cancer radiotherapy treatment requires a conceptual advance in radiobiology towards a more molecular biophysics approach. This will ultimately enable the development of novel molecular-based personalised treatment strategies into the future.

References

1. M. B. Barton, M. Frommer and J. Shafiq, *Lanc. Oncol.* **7**, 584 (2006).
2. D. E. Lea, *The Action of Radiation on Living Cells* (Cambridge University Press, London, 1946).
3. K. H. Chadwick and H. P. Leenhouts, *The Molecular Theory of Radiation Biology* (Springer, Berlin 1981).
4. H. Nikjoo, D. Emfietzoglou, R. Wanatabe and S. Uehara, *Rad. Phys. Chem.* **77**, 1270 (2008).
5. C. Champion, Quantum-mechanical contributions to numerical simulations of charged particle transport at the DNA scale, in *Radiation Damage in Biomolecular Systems*, Biological and Medical Physics, Biomedical Engineering (Springer, Netherlands, 2012), p. 263.
6. C. Champion, *Phys. Med. Biol.* **55**, 11 (2010).

7. Y. Gholami *et al.*, *J. Phys. Conf. Ser.* **489**, 012011 (2014).
8. H. Nikjoo and L. Lindborg, *Phys. Med. Biol.* **55**, R65 (2010).
9. L. D. Landau and E. P. Lifshitz, *Quantum Mechanics — Non-Relativistic Theory*, 3rd edn. (Oxford, Pergamon, Greece, 1984).
10. V. B. Berestetskii, L. P. Pitaevskii and E. P. Lifshitz, *Quantum Electrodynamics*, 2nd edn. (Butterworth-Heinmann, Oxford, 1982).
11. G. N. I. Clark *et al.*, *Mol. Phys.* **108**, 1415 (2010).
12. L. Van Hove, *Phys. Rev.* **95**, 249 (1954).
13. M. Michaud, A. Wen and L. Sanche, *Rad. Res.* **159**, 3 (2003).
14. N. Wanatabe, H. Hayashi and Y. Udagawa, *Bull. Chem. Soc. Jpn.* **70**, 719 (1997).
15. H. Hayashi, N. Wanatabe, Y. Udagawa and C.-C. Kao, *Proc. Nat. Acad. Sci. USA* **97**, 6264 (2000).
16. D. Emfietzoglou and H. Nikjoo, *Rad. Res.* **167**, 110 (2007).
17. C. Kittel, *Quantum Theory of Solids*, 2nd edn. (John Wiley & Sons, New York, 1963)
18. P. Pines and P. Nozieres, *The Theory of Quantum Liquids* (Perseus Books, Cambridge, 1966).
19. M. Dingfelder *et al.*, *Rad. Res.* **169**, 584 (2008).
20. M. Dingfelder, *Appl. Rad. Iso.* **83**, 142 (2014).
21. D. Emfietzoglou *et al.*, *Phys. Med. Biol.* **54**, 3451 (2009).
22. Z. Kuncic *et al.*, *Comp. Math. Meth. Med.* 147252 (2012).
23. W. R. Nelson, H. Hirayama and D. W. O. Rogers, *Report SLAC-265* (1985).
24. I. Kawrakow, *Med. Phys.* **27**, 485 (2000).
25. W. Friedland, M. Dingfelder, P. Kundrat and P. Jacob, *Mut. Res.* **711**, 28 (2011).
26. H. Nikjoo, S. Uehara, D. Emfietzoglou and F. A. Cucinotta, *Radiat. Meas.* **41**, 1052 (2006).
27. S. Agostinelli *et al.*, *Nuc. Instrum. Meth. Phys. Res. A* **506**, 250 (2003).
28. S. Incerti *et al.*, *Med. Phys.* **37**, 4692 (2010).
29. M. Karamitros, S. Incerti and C. Champion, *Rad. Onc.* **102**, S191 (2012).
30. M. Karamitros *et al.*, *Prog. Nuc. Sci. Tech.* **2**, 503 (2011).
31. Z. Francis, S. Incerti, M. Karamitros, H. N. Tran and C. Villagrasa, *Nuc. Inst. Meth. Phys. B* **269**, 2307 (2011).
32. I. Martinez-Rovira, J. Sempau and Y. Prezado, *Med. Phys.* **39**, 119 (2012).
33. J. Crosbie *et al.*, *Intl. J. Rad. Oncol. Biol. Phys.* **77**, 886 (2010).
34. K. Prise and J. M. O'Sullivan, *Nat. Rev. Canc.* **9**, 351 (2009).

Chapter 4

Continuous Time Random Walk and Migration–Proliferation Dichotomy of Brain Cancer

A. Iomin

Department of Physics, Technion
Haifa 32000, Israel
iomin@physics.technion.ac.il

A theory of fractional kinetics of glial cancer cells is presented. A role of the migration-proliferation dichotomy in the fractional cancer cell dynamics in the outer-invasive zone is discussed and explained in the framework of a continuous time random walk. The main suggested model is based on a construction of a 3D comb model, where the migration-proliferation dichotomy becomes naturally apparent and the outer-invasive zone of glioma cancer is considered as a fractal composite with a fractal dimension $D_{\mathrm{fr}} < 3$.

1. Introduction

Brain tumors result from the uncontrolled growth of abnormal cells, destruction of normal tissues, and invasion of vital organs. These processes can be subdivided into many types based on several classification characteristics and involve any of the cell types found in the brain, such as neurons, glial cells, astrocytes, or cells of the meninges.[1,2] The mechanisms behind cancer progression result from the accumulation of one or a few specific mutations that disrupt biological pathways like growth factor signaling, DNA damage repair, cell cycle, apoptosis and cellular adhesion.[3] Among all possible cancer cell genotypes, leading to six main alternations of malignant growth,[3] cell motility and invasion are the most important for the present consideration.

Glioma is one of the most recalcitrant brain disease, with an optimal therapy treatment survival period of 15 month and most tumors recur within 9 months of initial treatment.[4,5] One of the main possible reason of such devastating manifestation is the migration proliferation dichotomy of cancer cells. This phenomenon has been firstly observed at clinical investigations,[6,7] where it has been shown that in the outer invasive zone glioma cancer cells possesses a property of high motility, while the proliferation rate of these migratory cells is essentially lower than in the tumor core. This anti-correlation between proliferation and migration of cancer cells, also known as the Go or Grow hypothesis (see discussions in Refs. 8 and 9), suggests that cell division and cell migration are temporally exclusive phenotypes.[6]

The phenomenon that tumor cells defer proliferation for cell migration was also experimentally demonstrated[a] in Refs. 10–12. The switching process between these two phenotypes is still not well understood. Moreover, it should be mentioned that conflicting data appear in the literature concerning the Go or Grow hypothesis; details of discussions on this can be found in Refs. 8 and 9.

Extensive theoretical modelling follow this finding. Because a switching process between these two phenotypes still is not well understood, many efforts are directed to develop relevant models with relevant mechanisms of switching of the glioma cells, resulting in several phenomenological models. Comprehensive discussions of these models one can find in. Refs. 14–18. It was suggested by Khain et al.[14,19] that the motility of cancer cells is a function of their density. Multi-parametric modelling of the phenotype switching was considered in Ref. 20. The agent-based approach to simulate multi-scale glioma growth and invasion was used in Refs. 21 and 22. Subdiffusive cancer development on a comb was studied in Ref. 23, where a continuous time random walk (CTRW) was firstly suggested for metastatic cancer development. A stochastic approach for the proliferation-migration switching involving only two parameters was proposed in Refs. 24 and 25 where the transport of cancer cells was formulated in terms of the CTRW, as well. 'Go or Grow' mechanism was proposed in Ref. 15, where the transition to invasive tumor phenotypes can be explained on the basis of the oxygen shortage in the environment of a growing tumor. Phenotypic switching due to density effect was also suggested in Refs. 16 and 26. Both numerical and analytical approaches were developed in Ref. 17 to study the glioma propagation in the framework of reaction-diffusion equations, where the phenotype switching depends on oxygen in a threshold manner. Collective behavior of brain tumor cells under the hypoxia condition was studied in Ref. 27.

A new therapeutic method, recently suggested[28–30] for non-invasive treatment of glioma — brain cancer by a radio-frequency electric field, also opens new directions of understanding of glioma development. A specific task emerging here is whether this new medical technology is effective against invasive cells with a high motility, when a switching between migrating and proliferating phenotypes takes place. As is well known, one of the main features of malignant brain cancer is the ability of tumor cells to invade the normal tissue away from the multi-cell tumor core, causing treatment failure.[31] This problem relates to modelling of the dynamics of cancer glial cells in heterogeneous media (as brain cancer is) in the presence of a radio-frequency electric field, which acts as a tumor treating field (TTF).[28–30] As reported, this transcranial treatment by the low-intensity (1–3 V/cm), intermediate-frequency (100–200 kHz) alternating electric field, produced by electrode arrays applied to the scalp, destroys cancer cells that undergoing to division, while normal

---

[a]This kind of migration–proliferation dichotomy was also found at metastatic behavior of breast cancer.[13]

tissue cells are relatively not affected.[b] An important result of this new technology treatment is increasing the survival period in two.[30,c]

Therefore, an essential question is how the TTF affects aggressive migrating cells in the outer-invasive region with a low-rate of proliferation. To shed light on this situation, a simplified toy model for glioma treatment by the TTF has been suggested in Refs. 32 and 33, where a mathematical task of the migration–proliferation dichotomy was formulated in the CTRW framework.[34,35] Note, that the simplest mathematical realization of the CTRW mechanism of the migration–proliferation dichotomy was introduced for a comb model.[23] In the framework of this toy model, it was possible to estimate the effectiveness of the TTF treatment in the outer-invasive region of the tumor development.[32] It has also been shown that while the TTF is highly effective in the multi-cell tumor core, its action is ineffective in the presence of the migration–proliferation dichotomy.[32] This result is mainly based on the $1D$ consideration, where the fractal cancer composite is embedded in the $1D$ space. In reality, the situation is much more complicated, since the fractal cancer composite in the outer-invasive region develops in the $3D$ space. As a result of this, the TTF efficiency depends on the fractal dimension of the cancer composite in the outer-invasive region. Therefore, a more realistic model to estimate a medical effect of brain cancer (glioma) treatment by the RF electric field is suggested.[33] This model is based on a construction of a 3D comb model for the cancer cells, where the outer-invasive region of glioma cancer is considered as a fractal composite embedded in the 3D space. In the framework of this 3D model it was shown that the efficiency of the medical treatment by the TTF depends essentially on the mass fractal dimension D_{fr} of the cancer in the outer-invasive region.

In this paper we follow a CTRW consideration, suggested in Ref. 23. Description of fractional kinetics of glioma development under the TTF treatment in the framework of the one-dimensional (1D) and the three-dimensional (3D) comb models show that the efficiency of the medical treatment depends essentially on the mass fractal dimension of the cancer in the outer invasive zone.[32,33] The aim of this research is understanding both the role of the migration–proliferation dichotomy in fractional cancer cell transport and its influence on a therapeutic effect due to the TTF.

2. Self-Entrapping by Fission as Fractional Mechanism of Tumor Development

In this section we formulate the migration–proliferation dichotomy in the framework of the CTRW. A simplified scheme of cell dissemination through the vessel network

---

[b]An explanation of this phenomenon in the electrostatic framework is vague, since for this weak RF electric field of the order of $1\,\mathrm{V/cm}$, the inter-bridge voltage between the daughter cells is of the order $10^{-3}\,\mathrm{V}$ that is less than the voltage fluctuations related with the cell shape fluctuations.
[c]In the latest phase III trial study for TTF treatment for glioblastoma,[5] the TTF treatment has median survival of 6.6 months versus 6 months of the chemotherapy treatment.

was considered by means of the following two steps.[23,36] The first step is a biological process of cell fission. The duration of this stage is T_f. The second process is cell transport itself with duration T_t. Therefore the cell dissemination is approximately characterized by the fission time T_f and the transport time T_t. During the time scale T_f, the cells interact strongly with the environment and motility of the cells is vanishingly small. The duration of T_f could be arbitrarily large. During the second time T_t, interaction between the cells is weak and motility of the cells leads to cell invasion, which is a very complex process controlled by matrix adhesion.[6] It involves several steps including receptor-mediated adhesion of cells to extracellular matrix (ECM), matrix degradation by tumor-secreted proteases (proteolysis), detachment from ECM adhesion sites, and active invasion into intercellular space created by protease degradation. It is convenient to introduce a "jump" length X_t of these detachments as a distance which a cell travels during the time T_t. Hence, the cells form an initial packet of free spreading particles, and the contribution of cell dissemination to the tumor development process consists of the following time consequences:

$$T_f(1)T_t(2)T_f(3)\ldots. \tag{1}$$

There are different realizations of this chain of times, due to different durations of $T_f(i)$ and $T_t(i)$, where $i = 1, 2, \ldots$. Therefore, one concludes that transport is characterized by random values $T(i)$ which are waiting (or self-entrapping) times between any two successive jumps of random length $X(i)$. This phenomenon is known as a continuous time random walk (CTRW).[37] It arises as a result of a sequence of independent identically distributed random waiting times $T(i)$, each having the same probability density function (PDF) $w(t)$, $t > 0$ with a mean characteristic time T and a sequence of independent identically distributed random jumps, $x = X(i)$, each having the same PDF $\lambda(x)$ with a jump length variance σ^2. It is worth mentioning that a cell carries its own trap, by which it is set apart from transport. This process of self-entrapping differs from the standard CTRW, where traps are external with respect to the transporting particles. The crucial point of the fractional transport is the power law behavior of the waiting time PDF

$$w(t) = \alpha T/(1 + t/T)^{1+\alpha} \tag{2}$$

where $0 < \alpha < 1$ and T is a characteristic time. In this case the averaged time is infinite. A proper explanation of Eq. (2) can be the following quotation from Ref. 34: "A process with the long-tailed pausing time distribution would suffer a very sporadic behavior — long intermittencies may exist, followed by bursts of events. The more probable pauses between events would be short but occasionally very long pauses would exist. Given a long pause, there is still a smaller but finite probability that an even longer one will occur. It is on this basis that one would not be able to measure a mean pausing time by examining data." Some justification of Eq. (2) for the fission times can be presented by proposing multi-time scales of self-entrapping. We can consider that self-entrapping for different generations of cells has different mean characteristic time scales, see Appendix A. One obtains that the PDF, which accounts for all exit events from proliferation occurring on all time scales, has the power law asymptotic of Eq. (2). The obtained distribution of Eq. (2) is valid, when cell transport is considered on a fractional subdiffusive structure such as a comb model.

3. Comb-Like Model with Proliferation

Fractional transport of cells, namely subdiffusion, can be described in the framework of the comb model.[38] The comb model is an example of subdiffusive 1D media where CTRW takes place along the x structure axis. Diffusion in the y direction plays the role of traps with the PDF of delay times of the form $w(t) \sim 1/(1 + t/\mathcal{T})^{3/2}$. A special behavior of diffusion on the comb structure is that the displacement in the x direction is possible only along the structure axis (x-axis at $y = 0$). Thus, the diffusion coefficient in the x direction is $D_{xx} = D\delta(y)$, while the diffusion coefficient in the transversal y direction is a constant $D_{yy} = D_0$. A random walk on the comb structure is described by the distribution function $P = P(x, y, t)$ and the current

$$\mathbf{j} = \left(-\delta(y)D\frac{\partial P}{\partial x}, -D_0\frac{\partial P}{\partial y} \right).$$

The continuity equation with proliferation $C(P)$ yields the following Fokker–Planck equation

$$\frac{\partial P}{\partial t} - \delta(y)D\frac{\partial^2 P}{\partial x^2} - D_0\frac{\partial^2 P}{\partial y^2} = C(P), \tag{4}$$

where the diffusion coefficients can be related to the CTRW parameters $D = \sigma^3/\mathcal{T}$. The initial condition $P(x, y, 0) = P_0(x)\delta(y)$ is an initial distribution on the x-axis, and the boundary conditions are taken on infinities $P(t) = P'(t) = 0$ for both the x- and y-coordinates. The primes denote the spatial derivatives.

It is convenient to work with dimensionless variables and parameters. In the case of normal diffusion, when $D_x = $ const, the dimensionless time and coordinates are obtained by rescaling with relevant combinations of the comb parameters D_x and D_0. One obtains the following dimension variables for time $(D_0^3/D_x^2)t \to t$ and for the coordinates $D_0x/D_x \to x$, $D_0y/D_x \to y$.

We consider a possible mechanism of tumor cell proliferation. The term $C(P)$ in Eq. (4) determines the change in the total number of transporting cells due to proliferation at rate $\tilde{C}$. This can be considered as a linear approximation of a logistic population growth[39]

$$C(P) = \tilde{C}P(1 - P/K), \tag{5}$$

where K is the carrying capacity of the environment (see e.g. Ref. 40). It is worth stressing that linearization is important in the use of the powerful machinery of the Laplace transform. When $P/K \to P < 1/2$ and $\mathcal{C} = K\tilde{\mathcal{C}}D_x^2/D_0^3$, then the linearization $C(P) = \mathcal{C}P$ is valid.[39] In the opposite case, when $P > 1/2$ the growth is approximated by $C(P) = \mathcal{C}\bar{P}$, where $\bar{P} = 1 - P$. According to the migration–proliferation dichotomy in the comb model, the transporting cells along the x axis do not proliferate. This means that cells proliferate only if they have a non-zero y coordinate. Therefore, $C(P) = \mathcal{C}(1 - \delta(y))P$, and Eq. (4) reads in the dimensionless form

$$\frac{\partial P}{\partial t} - \delta(y)\frac{\partial^2 P}{\partial x^2} - \frac{\partial^2 P}{\partial y^2} = \mathcal{C}\left(1 - \delta(y)\right)P. \tag{6}$$

When $\mathcal{C} > 0$, Eq. (6) describes cell transport with proliferation, and the PDF P corresponds to a low concentration of cells. In the opposite case, when $\mathcal{C} < 0$, Eq. (6) describes fractional cell transport with degradation that corresponds to a high cell concentration, and P exchanges for $\bar{P}$.

The first term in the r.h.s. of Eq. (6) is eliminated by substitution $P = e^{\mathcal{C}t}F$. Carrying out the Laplace transform $\tilde{F}(s, x, y) = \hat{\mathcal{L}}[F(x, y, t)]$ and looking for the solution in the form $\tilde{F} = e^{-\sqrt{s}|y|}f(x, s)$, one obtains

$$F(x, y, t) = \hat{\mathcal{L}}^{-1}[f(x, s)\exp(-\sqrt{s}|y|)]. \tag{7}$$

As admitted, the true motion is in the x-axis, while the y-axis is an auxiliary, and integration over y is performed. Integrating Eq. (6) with respect to the variable y and introducing the PDF

$$P_1(x, t) = \int_{-\infty}^{\infty} P(x, y, t)dy, \tag{8}$$

one obtains the following equation for $F_1 = e^{-\mathcal{C}t}P_1$ in the Laplace space $\tilde{F}_1(s) = \hat{\mathcal{L}}[F_1(t)]$:

$$s\tilde{F}_1 - \partial_x^2 f = P_0(x) - \mathcal{C}f. \tag{9}$$

Integrating Eq. (7) over y, we obtain a relation between the PDFs of the total number of cells F_1 and transporting number of cells f in the Laplace space

$$f \equiv \tilde{F}(x, y = 0, s) = (1/2)\sqrt{s}\tilde{F}_1(x, s).$$

Substitution of this relation in Eq. (9) yields, after the Laplace inversion, the Fokker–Planck equation for the distribution F_1. To this end, Eq. (9) is multiplied by $\sqrt{s}$ and then by virtue of Eq. (C.6) the inverse Laplace transform yields the following equation for F_1

$$2D_{\mathcal{C}}^{1/2}F_1 - \partial_x^2 F_1 = -\mathcal{C}F_1, \tag{11}$$

where $D_{\mathcal{C}}^\alpha$ is the fractional derivative in the Caputo form[41,42] (see Appendix C). This equation describes fractional transport of cells with fission when $\mathcal{C} > 0$ and degradation when $\mathcal{C} < 0$, where the sign of $\mathcal{C}$ depends on either $P = e^{\mathcal{C}t}F < 1/2$, or $P > 1/2$.[d]

4. Fractional Dynamics of Untreated Cancer

As shown, the cell fission is a source of the fractional time derivatives. This equation can be extended for an arbitrary fractional exponent $0 < \alpha < 1$ and $1/2 \to \alpha$. Therefore, this generalization of Eq. (11) yields

$$D_{\mathcal{C}}^{\alpha} F_1 - \alpha \partial_x^2 F_1 = -\alpha \mathcal{C} F_1. \tag{12}$$

Taking into account that $D_{\mathcal{C}}^{\alpha}$ can be expressed by the Riemann–Liouville fractional derivatives D_{RL}^{α} (see Appendix C) $D_{\mathcal{C}}^{\alpha} = D_{RL}^{\alpha-1} D_{RL}^1$, we obtain another standard form for the fractional Fokker–Planck equation (FFPE) with proliferation, or degradation,

$$\frac{\partial F_1}{\partial t} - \alpha D_{RL}^{1-\alpha} \frac{\partial^2 F_1}{\partial x^2} = -\alpha \mathcal{C} D_{RL}^{1-\alpha} F_1. \tag{13}$$

To solve Eq. (13), we use the separation of variables.[35] We consider an analytical solution for the $P < 1/2$ using the following substitution

$$F_1(x, t) = \sum_n T_n(t) \phi_n(x). \tag{14}$$

Therefore, a solution which corresponds to the initial condition $P_0(x)$, is determined by the Green function $G(x, t | x', 0)$:

$$F_1(x, t) = \int_{-\infty}^{\infty} dx' G(x, t | x', 0) P_0(x')$$

$$= \int_{-\infty}^{\infty} dx' \int dk T_k(t) \phi_k(x) \phi_k^*(x') P_0(x'). \tag{15}$$

Here $\phi_k(x)$ is a solution of the eigenvalue problem

$$-\frac{\partial^2 \phi_k}{\partial x^2} = \lambda(k) \phi_k,$$

where $\lambda(k) = k^2$ is the continuous spectrum with eigenfunctions

$$\phi_k(x) = \exp[\pm kx]. \tag{16}$$

The temporal eigenfunction $T_k(t)$ is governed by the fractional equation

$$\dot{T}_k(t) + \alpha \lambda_{\mathcal{C}}(k) D_{RL}^{1-\alpha} T_k(t) = 0, \tag{17}$$

---

[d]Since $\partial \bar{P} = -\partial P$, Eq. (6) for P (when $P < 1/2$) just coincides with one for $\bar{P} = 1 - P$ (when $P > 1/2$). The only difference is when $P < 1/2$, $\mathcal{C} > 0$, while for $P > 1/2$ one has $\mathcal{C} < 0$.

where $\lambda_{\mathcal{C}}(k) = (k^2 + \mathcal{C})$. The solution is described by the Mittag–Leffler function[43] $E_\alpha(z) \equiv E_{\alpha,1}(z)$ (see Appendix C)

$$T_k(t) = E_\alpha[\alpha\lambda_{\mathcal{C}}(k)t^\alpha], \tag{18}$$

where $T_k(0) = 1$, and $E_\alpha(z)$ has the initial stretched exponent behavior

$$T_k(t) \sim \exp[-[\alpha\lambda_{\mathcal{C}}(k)t^\alpha/\Gamma(1+\alpha)] \tag{19}$$

which turns over to the power law long-time asymptotics

$$T_k(t) \sim [\Gamma(1-\alpha)\alpha\lambda_{\mathcal{C}}(k)t^\alpha]^{-1}. \tag{20}$$

Using these properties of $E_\alpha(z)$, the fractional spreading of cancer cells can be evaluated analytically for both initial and long-time behaviors. Substitution of Eqs. (16) and (19) in Eq. (15) yields the following initial time solution

$$P_1(x,t) \propto \sqrt{\frac{\pi\Gamma(1+\alpha)}{\alpha t^\alpha}} \exp[\mathcal{C}t - \alpha\mathcal{C}t^\alpha/\Gamma(1+\alpha)]$$
$$\times \exp[-\Gamma(1+\alpha)x^2/4\alpha t^\alpha]. \tag{21}$$

Analogously, the long-time solution is

$$P_1(x,t) \propto \frac{1}{\alpha t^\alpha\Gamma(1-\alpha)} \exp[\mathcal{C}t - \sqrt{\mathcal{C}}|x|], \tag{22}$$

where we take, for clarity, $P_0 = \delta(x)$ for both the short and long time solutions. These two solutions (21) and (22) corresponds to different scales. Solution of Eq. (22) describes long-time dynamics. When the argument in the exponential function is zero, it corresponds to the front of cell invasion with equation $x \sim l_0 = \sqrt{\mathcal{C}}t$. This is a so-called linear model which describes a solid tumor growth. In this region with $x < l_0$ the exponential growth $e^{\mathcal{C}t}$ is dominant. Subdiffusion described by Eq. (21) corresponds to the cell transport in the outer-invasive zone with $x > l_0$. When $\mathcal{C} \to 0$ only this solution takes place. Therefore, the cell spreading in the core region with $x < l_0$ is due to the cell proliferation, while in the outer-invasive zone the cell motility is the main engine of the cell spreading.

5. Cell Kinetics in Presence of the TTF

Let us consider cell kinetics in the outer-invasive zone in more detail. To this end we consider the fractal cancer development in the presence of the TTF. This process can be described by fractional kinetics in the framework of the comb model, as well, where it is easier to draw an intelligible picture of interplay between high-motility of aggressive cancer cells and the TTF in the outer-invasive region. Contrary to the $1D$ comb model, in this section we extend our consideration of the treated cancer to the three dimensional cancer development, where proliferation takes place inside a fractal composite, embedded in the $3D$ space with the fractal dimension $D_{\mathrm{fr}} < 3$.

In the $3D$ comb model, this anomalous diffusion can be described by the $4D$ distribution function $P = P(\mathbf{x}, y, t)$, and by analogy with the $1D$ comb model (6), a special behavior here is the displacement in the $3D$ x space at $y = 0$. The Fokker–Planck equation in the same dimensionless variables reads

$$\partial_t P = \delta(y)\Delta P + d\partial_y^2 P, \tag{23}$$

where d is an effective diffusion coefficient and $\Delta = \sum_{j=1}^{3} \partial_{x_j}^2$.

5.1. *Comb model with proliferation and TTF*

Obviously, cell fission/division is random in the x space and discontinuous, contrary to that in the tumor core. Therefore, the outer-invasive region of the cancer can be reasonably considered as a random fractal set $F_{D_{\mathrm{fr}}}(\mathbf{x}) = F_\alpha(x_1) \times F_\beta(x_2) \times F_\gamma(x_3)$ embedded in the $3D$ space, for example, as for low-grade astrocytomas,[7] with the fractal dimension, $0 < D_{\mathrm{fr}} < 3$ and $\alpha + \beta + \gamma = D_{\mathrm{fr}}$. For simplicity, we take $\alpha = \beta = \gamma = D_{\mathrm{fr}}/3 = \nu$.

The effective diffusion coefficient in Eq. (23) becomes inhomogeneous $d \rightarrow d\chi(\mathbf{x})$, where $\chi(\mathbf{x}) = \chi(x_1)\chi(x_2)\chi(x_3)$ is a characteristic function of the fractal, such that $\chi(x_j) = 1$ for $x_j \in F_{D_{\mathrm{fr}}}(\mathbf{x})$ and $\chi(x_j) = 0$ for $x_j \notin F_{D_{\mathrm{fr}}}(\mathbf{x})$, where x_j are the Cartesian coordinates $j = 1, 2, 3$. Now we take into account the influence of the TTF that affects (destroys) only quiescent cells, belonged to the proliferation phenotype, according to Refs. 28–30. Mathematically, this process is expressed by diffusion in the y direction with decay:

$$d\frac{\partial^2 P(\mathbf{x}, y, t)}{\partial y^2} \Rightarrow \chi(\mathbf{x}) \left[d\frac{\partial^2}{\partial y^2} - C \right] P(\mathbf{x}, y, t), \tag{24}$$

where coefficient C defines a difference between the proliferation and the degradation rate. In general case, C is a random function of time and space. For example, it was considered as a random death rate for the random walk in the discrete inhomogeneous media.[44] Here we take it as a positive averaged constant value.

Summarizing these arguments, mapping the glioma problem onto the $4D$ comb model can be described by the following rules: (i) The dynamics of cancer cells takes place in the $3D$ space, which is described by three x-coordinates (x_1, x_2, x_3). (ii) The y-axis corresponds to a supplementary coordinate that introduces the migration–proliferation dichotomy for the model. Therefore, at $y = 0$ the cells migrates and are not affected by the TTF. Contrarily, the cells with $y \neq 0$ proliferate and are subjected to the TTF.

Taking this into account, one arrives at the equation of the cancer development in the presence of the TTF

$$\frac{\partial P}{\partial t} = \delta(y)\Delta P + \chi(\mathbf{x}) \left[d\frac{\partial^2}{\partial y^2} - C \right] P(\mathbf{x}, y, t). \tag{25}$$

First, we apply the Fourier transform to Eq. (25) with respect to the x_j coordinates. To this end, we rewrite Eq. (25) in the form of convolution integrals.

Therefore, as shown in Ref. 33 and in Appendix B, fractal cancer development in the presence of the TTF can be considered as a random fractal composite of cancer cells embedded in the 3D. Following coarse graining and averaging procedure, described in Appendix B, we arrive at the 3D comb model that describes the fractal cancer development in the outer-invasive region of glioma in the presence of the TTF

$$\partial_t P(r,y,t) = \delta(y)\Delta P(r,y,t) + [d\partial_y^2 - C](-\Delta)^{\frac{3-D_{\text{fr}}}{2}} P(r,y,t), \tag{26}$$

where $P(r,y,t)$ is the radial function in the 3D x space $r = |\mathbf{x}|$.

5.2. Dynamics in the Fourier-Laplace space

Equation (26) can be considered as a starting point of the analysis, and its solution will be obtained by means of the Fourier and the Laplace transforms. Performing the Fourier transform, constructed in the Appendix B, one obtains Eq. (26) in the Fourier space

$$\partial_t \bar{P} = -k^2 \delta(y)\hat{P} + k^{3-D_{\text{fr}}}[d\partial_y^2 \hat{P} - C\bar{P}]. \tag{27}$$

The last term in the r.h.s. of Eq. (27) is eliminated by the substitution

$$\bar{P}(k,y,t) = \exp(-Ck^{3-D_{\text{fr}}}t)F(k,y,t) = e^{-Ck^{-\alpha}t}F(k,y,t), \tag{28}$$

where $\alpha = D_{\text{fr}} - 3$.

The next step of the analysis is the Laplace transform in the time domain

$$\hat{\mathcal{L}}[F(k,y,t)] = \tilde{F}(k,y,s).$$

Looking for the solution of the Laplace image in the form

$$\tilde{F}(k,y,s) = \exp[-|y|\sqrt{k^\alpha s/d}]f(k,s), \tag{29}$$

one arrives at the intermediate expression in the form of the Laplace and Fourier inversions

$$P(r,y,t) = \hat{\mathcal{F}}_k^{-1}\left\{\exp(-Ck^{-\alpha}t)\hat{\mathcal{L}}_t^{-1}\left[\frac{e^{-|y|\sqrt{sk^\alpha/d}}}{2\sqrt{sdk^{-\alpha}} + k^2}\right]\right\}. \tag{30}$$

As admitted above, the y-axis is the auxiliary, or supplementary coordinate, which determines the cell proliferating process (cell fission). Therefore to find the complete distribution of cancer cells in the x space, integration over y is performed (see Sec. 4):

$$\overline{P}(r,t) = \int_{-\infty}^{\infty} P(r,y,t)dy. \tag{31}$$

Both the integration over y and the inverse Laplace transform are carried out exactly. This, eventually, yields a solution in the form the $3D$ Fourier inversion

$$\overline{P}(r,t) = \frac{1}{(2\pi)^3} \int_{-\infty}^{\infty} e^{-i\mathbf{k}\cdot\mathbf{x}} \exp(-Ck^{3-D_{\text{fr}}}t)\mathcal{E}_{\frac{1}{2}}\left(-\frac{1}{2}\sqrt{k^{1+D_{\text{fr}}}t/d}\right) d^3k. \quad (32)$$

Here

$$\mathcal{E}_\alpha(-z) = \frac{1}{2\pi i} \int_\gamma \frac{u^{\alpha-1}e^u du}{u^\alpha + z}$$

is the Mittag–Leffler function defined by the inverse Laplace transform with a corresponding deformation of the contour of the integration.[43]

5.3. *True distributions*

Solution (32) is a convolution of the kernel of the TTF treatment $\mathcal{R}(z)$ and the untreated cancer distribution $\mathcal{P}(z)$

$$\overline{P}(r,t) = \mathcal{R} \star \mathcal{P}. \quad (33)$$

When $C = 0$, which means that the TTF compensates proliferation, the solution is described by the Mittag–Leffler function with the scaling variable $z = r/t^{\frac{1}{1+D_{\text{fr}}}}$. This scaling determines the cancer cell expansion

$$r \sim t^{\frac{1}{1+D_{\text{fr}}}} \quad (34)$$

that depends essentially on the fractal dimension of the proliferation volume of the fractal cancer composite and reflects the migration–proliferation dichotomy in the outer-invasive region. Indeed, for the fractal cancer volume (or mass) $\mu(r) \sim r^{D_{\text{fr}}}$, the cancer development is superdiffusive when $D_{\text{fr}} < 1$, while for $D_{\text{fr}} > 1$ the latter spreads subdiffusively. This property is pure kinetic and, apparently, is universal for the cancer development and related to the fractal dimension of the cancer.

Now let us return to the convolution integral (33). To avoid awkward expressions of integrations with the hypergeometric functions, we consider particular cases of the fractal dimension $D_{\text{fr}} = 2$ and $D_{\text{fr}} = 1$. For $D_{\text{fr}} = 2$, due to the scaling argument, one obtains that the untreated cancer spreads subdiffusively $\langle r^2 \rangle \sim t^{\frac{2}{3}}$, while for the TTF kernel we have

$$\mathcal{R}(r,t) = \frac{1}{\sqrt{(2\pi)^3 r}} \int_0^\infty e^{-Ctk} k^{\frac{3}{2}} J_{\frac{1}{2}}(kr)dk$$

$$= \frac{1}{3\pi^2(Ct)^3} {}_2F_1\left(\frac{3}{2}, 2; \frac{3}{2}; -\frac{r^2}{(Ct)^2}\right). \quad (35)$$

Here ${}_2F_1(a, b; c; z)$ is the hypergeometric function. This yields the power law decay of the distribution function

$$\mathcal{R}(r,t) = \frac{(Ct)^{-3}\pi^{-2}}{3[1 + r^2/(Ct)^2]^2}. \quad (36)$$

This power law kernel shows that the TTF is inefficient for $D_{\text{fr}} > 1$ in the presence of the migration proliferation dichotomy. It is tempting to calculate the second moment $\langle r^2(t) \rangle$ with the distribution $\overline{P}(r,t)$. In this case, one should recognize that a cutoff $r = t$ of the Lévy flights for $r, t \gg 1$ should be performed. This is a well known procedure,[45] used for the Lévy walks.

5.3.1. $D_{fr} = 1$

The situation changes when $D_{fr} \leq 1$. In this case the TTF leads to the Brown exponential cutoff of the cancer spread in Eq. (33). For $D_{fr} = 1$ the problem is analytically treatable. For the small argument, which corresponds (for a short time) to a long-scale tail of the distribution, the Mittag–Leffler function decays exponentially[35,43] $\exp(-K_{\frac{1}{2}} \sqrt{|k|^{1+D_{fr}}t})$ with the generalized transport coefficient $K_{\frac{1}{2}} = [2\Gamma(3/2)\sqrt{d}]^{-1}$. This yields the solution for the compensated cancer with $C = 0$ in the form of the hypergeometric functions like in Eqs. (35) and (36). Following,[46] one obtains

$$\mathcal{P}(r,t) = \frac{K_{\frac{1}{2}}\sqrt{t}}{(2\pi)^3(1 + r^2/K_{\frac{1}{2}}^2 t)^2} \left[K_{\frac{1}{2}}\sqrt{t} + \sqrt{K_{\frac{1}{2}}^2 t + r^2} \right]^{-\frac{1}{2}}. \tag{37}$$

This metastatic power law behavior is restricted by the Brown distribution due the TTF kernel

$$\mathcal{R}(r,t) = \hat{\mathcal{F}}^{-1}[e^{-Ck^2 t}] = \frac{1}{(4\pi Ct)^{3/2}} \exp\left(-\frac{r^2}{4Ct}\right). \tag{38}$$

The second moment is a good characteristic to show the TTF influence. One obtains from Eq. (37) for the compensated cancer $\langle r^2 \rangle \sim t^{\frac{3}{2}}$ for $r \gg 1$ that corresponds to superdiffusion at the large scale asymptotics, and the cutoff at $r = t$ is taken into account. The same calculation with the TTF kernel yields an effective treatment with $\langle r^2 \rangle \sim t^{\frac{3}{4}}$ that corresponds to the superdiffusion–subdiffusion transition due to the TTF. Obviously, that untreated cancer with $C < 0$ leads to the exponential spreading of cancer cells due to the exponential proliferation.

5.4. Numerical estimations of Eq. (32)

As shown in Eqs. (36) and (38) analytical form of the TTF operator depends on fractal dimension D_{fr}. Since analytical estimation of Eq. (32) leads to awkward expressions of integrations with the hypergeometric functions, numerical procedure is performed. The results are depicted in Fig. 1 for different values of the fractal dimension D_{fr}.

As obtained, the maximal therapeutic effect takes place at $D_{fr} = 3$ that immediately follows from Eq. (32). One should recognize that the solution for the compensated cancer $\mathcal{P}(r,t)$ is exactly the form of the interplay between the TTF and subdiffusion, which is the result of the migration–proliferation dichotomy. Another manifestation of this interplay is the fractal dimension $D_{fr} < 3$ that leads to metastatic behavior of either the TTF kernel or compensated cancer solution.

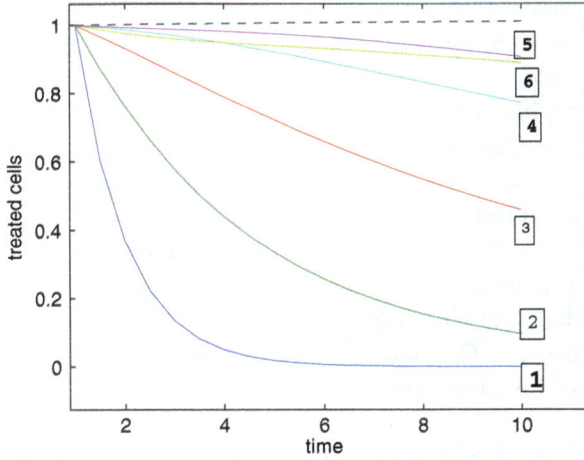

Fig. 1. (Color online) Dynamics of the treated cancer distribution for different values of the fractal dimension D_{fr}, where from plots from 1 to 6 correspond to $D_{\mathrm{fr}} = (3, 2.5, 2, 1.5, 1, 0.5)$.

6. Conclusion

The present study focuses on the influence of cell proliferation on transport properties. The mathematical formulation of this proliferation-migration dichotomy is based on the two main stages: cell fission with the self-entrapping time $\mathcal{T}_f$ and cell transport with durations $\mathcal{T}_t$. By virtue of these two time scales a description of tumor development is reduced to a CTRW process. A toy model of cancer development is suggested by using heuristic arguments on the relation between tumor development and the CTRW. In this case a fractional tumor development becomes a well defined problem since a mathematical apparatus of CTRW is well established (see e.g. Refs. 35, 34, 41 and 47). The constructed model is a modification of a so-called comb structure.[38] An important feature of this consideration of cell transport in the framework of the comb model is an essential enhancement of anomalous transport due to proliferation. Moreover, we obtained that the distribution function of the fractional transport depends on only two parameters, namely, scaled proliferation rate $\mathcal{C}$ and the fractional exponent α, where $\alpha = 1/2$ for the comb model.

The next step is studying glioma cancer development in the outer-invasive region in the presence of a tumor treating field. The model is based on a construction of a 3D comb model for the cancer cell transport, where the outer-invasive region of glioma cancer is considered as a fractal composite, embedded in the 3D host of the normal cell tissues. The description is performed in the four-dimensional $(\mathbf{x}, y)$ space, where the real three-dimensional $\mathbf{x}$ space stands for the description of real cancer development, while the supplementary y-coordinate is introduced to described a non-Markovian process in the framework of the Markovian description. From the biological point of view, this corresponds to the migration–proliferation dichotomy of the cancer cells, where the influence of the TTF is considered, as well. Therefore, the kinetic equation (26) in the $(\mathbf{x}, y)$ space is constructed by means of the coarse-graining, or embedding procedure inside the fractal space. This corresponds to the averaging in the 3D Fourier space in Eq. (B.6), and can be (roughly) considered as a generalization of the $1D$ procedure, based on averaging extensive physical values and expressed by means of a smooth function over a Cantor set that, eventually, leads to fractional integration.[48,49]

The efficiency of the TTF is estimated in the form of the convolution in Eqs. (32) and (33). This expressions describe the influence of the TTF on the cancer development. The efficiency of the medical treatment by the TTF depends essentially on the fractal dimension D_{fr}, and the TTF is the most efficient for $D_{\mathrm{fr}} = 3$. But, in reality, the outer-invasive zone is a fractal composite with the fractal dimension $D_{\mathrm{fr}} < 3$.

Another result relates to the spread of the compensated cancer, which is determined by Eq. (34). This result reflects the migration–proliferation dichotomy, namely the dependence of the cancer cells spread on the fractal dimension of the proliferation volume. For example, the cancer development is superdiffusive when $D_{\mathrm{fr}} < 1$, while for $D_{\mathrm{fr}} > 1$ it spreads subdiffusively. This property is pure kinetic and, apparently, is universal for the cancer development with a variety of biochemical processes. Recently an experimental validation of this kind phenomenon has been obtained for the metastatic detaching under *in vitro* studying the breast cancer.[13] Important question here also is aging the treatment. Initial times of the cancer development and treatment are different, and the time difference is unknown. Moreover, since the TTF acts on the proliferating cells only, the migration proliferation dichotomy leads to a particular case of a more general problem of aging population splitting.[50] For the present analysis this general approach[50,51] can be important for understanding the efficiency of the TTF. This can be an interesting issue for future studies.

It is also worth noting that a general solution in the form of the convolution $\overline{P}(r,t) = \mathcal{R} \star \mathcal{P}$ makes it possible to consider the compensated cancer in a more general framework of the fractional Fokker-Planck equation, namely[52]

$$\partial_t^p \mathcal{P} = (-\Delta)^{\frac{q}{2}} \mathcal{P},$$

where Caputo fractional derivative ∂_t^p is responsible for the migration–proliferation dichotomy, while $q = q(D_{\mathrm{fr}})$ reflects the fractal dimension of the tumor development in the outer-invasive region. In our case, $p = \frac{1}{2}$, according the comb model construction, while $q = (1 + D_{\mathrm{fr}})/4$ is an universal parameter, which determines the fractional space derivative due to the fractal dimension of the quiescent/proliferating cancer cells. In general case, $p \in (0, 1)$, and it is determined by another way of the introduction of the supplementary variable y.[44,53]

In conclusion, we discuss briefly a possible direct experiment, confirming the existence of a fractal cancer composite in the outer-invasive region. Cancer was considered as a fractal composite where a random fractal inclusion of the cancer cells $F_{D_{\mathrm{fr}}}(\mathbf{x}) = F_\alpha(x_1) \times F_\beta(x_2) \times F_\gamma(x_3)$ is embedded in the $3D$ space of normal tissue cells. Therefore, in the presence of the TTF, one can consider the frequency-dependent permittivities of migrating cancer cells ε_m and the normal tissue cells ε_n.[32] Under certain frequency of the TTF, the condition $\varepsilon_m < \varepsilon_n$ can be fulfilled. These permittivities were observed in time domain dielectric spectroscopy in experimental studies of the static and dynamic dielectric properties of normal, transformed, and malignant B- and T-lymphocytes.[57] The solution of the Maxwell equation for the electrostatic field in the frequency domain yields an essential enhancement of the respond field inside the random fractal dielectric composite of cancer cells. Therefore, the respond electric field can be large enough to break the cell membrane. For example, as shown in Ref. 32 the electric field response can be of the order of $10^4 \div 10^5$ V/cm, which exerts the irreversible electroporation[58] due to the external TTF with amplitude ~ 1 V/cm. This can be a mechanism for ablation of cancer cells, which effectively acts on migratory cancer cells.

A key quantity of this cancer treatment is localization of the electroporation field inside the cancer. There is a straightforward analogy with nanoplasmonics (see e.g. Refs. 59 and 60), where the electric field enhancement is due to a so-called surface plasmon resonance for a metal-dielectric composite, and localized surface plasmon oscillations are charge density oscillations confined to the conducting fractal nanostructure. The essential difference is that this biological cell enhancement of the electric field is not resonant, but geometrical due to the fractal cancer structure.[61] In this connection, *in vitro* experiments can be important for further understanding the interplay between the TTF and the migration–proliferation dichotomy. Theoretical description of this phenomenon needs more realistic assumptions than those suggested here in the framework of the comb model. Such studies should be performed in the framework of more sophisticated models of the switching between the migration and proliferation phenotypes,[14–18,24,25] and the next step is understanding how dielectric properties of cells correlate with cell motility and cell fission. Such experimental studies of the glioma cells can not be overestimated.

Acknowledgments

This work was supported by the Israel Science Foundation (ISF).

Appendix A. Power Law PDF

As an example, we consider that j-th generation of self-entrapping is the Poisson process

$$w_j(t) = \tau_j^{-1} \exp(-t/\tau_j)$$

with the characteristic time scale $\tau_j = \tau^j$, where $\tau = \tau_1 = \mathcal{T}$ is now an average time of cell divisions for the first generation. Therefore, following[54,55] and repeating exactly the analysis of Ref. 55, we obtain, by taking into account events occurring on all time scales, the following distribution:

$$w(t) = \frac{1-b}{b} \sum_{j=1}^{\infty} b^j \tau^{-j} \exp(-t/\tau^l),$$

where $b < 1$ is a normalization constant. Therefore, the last expression is a normalized sum and

$$w(t/\tau) = \tau w(t)/b - (1-b)\exp(-t/\tau)/b.$$

Using conditions $t \gg \tau > 1/b$, one obtains that at longer times $w(t/\tau) = \tau w(t)/b$. The last expression is equivalent to

$$w(t) \sim 1/t^{1+\alpha}, \tag{A.1}$$

where $\alpha = \ln(b)/\ln(1/\tau)$.

Appendix B. Coarse-Graining Procedure of Fractal Cancer Composite

Using the auxiliary identity

$$\chi(x_j)f(x_j) \equiv \partial_{x_j} \int_{-\infty}^{x_j} \chi(y)f(y)dy \equiv -\partial_{x_j} \int_{x_j}^{\infty} \chi(y)f(y)dy$$

with the boundary conditions $P(x_i = \pm\infty) = 0$, this integration with the characteristic function can be carried out by means of a convolution.[52,56,62] Note, that

$$\int_{-\infty}^{\infty} \chi(y)f(y)dy = \sum_{x_j \in F_\nu} \int_{-\infty}^{\infty} f(y)\delta(y - x_j)dy,$$

where $\sum_{x_j \in F_\nu} \delta(y - x_j) = \mu'(x) \sim |x|^{\nu-1}$ is a fractal density, such that on the finite interval $(-x, x)$, the integral

$$\int_{-x}^{x} d\mu(y) \sim |x|^\nu$$

corresponds to the fractal volume. Therefore, due to Theorem 3.1 in Ref. 62 we have

$$\int_0^x f(y)d\mu(y) \simeq \frac{1}{\Gamma(\nu)} \int_0^x (x-y)^{\nu-1} f(y)dy,$$

which is defined for the finite fractal volume $\mu(x) \equiv \mu(x_j)$.

   In what follows we will use the terminology and useful notations of fractional integration and differentiation.[35,41,42,47] Fractional integration of the order of ν is defined by the operator (see Appendix C)

$$_{-\infty}I_x^\nu f(x) = \frac{1}{\Gamma(\nu)} \int_{\infty}^x f(y)(x-y)^{\nu-1}dy, \tag{B.1}$$

$$_xI_\infty^\nu f(x) = \frac{1}{\Gamma(\nu)} \int_x^\infty f(y)(y-x)^{\nu-1}dy, \tag{B.2}$$

where $0 < \nu < 1$ and $\Gamma(\nu)$ is the Gamma function. By means of these fractional integration and differentiation (see Appendix C)

$$\mathcal{W}_-^{1-\nu} f(x) = \partial_x[_{-\infty} I_x^\nu f(x)], \qquad (B.3)$$

$$\mathcal{W}_+^{1-\nu} f(x) = \partial_x[_x I_\infty^\nu f(x)], \qquad (B.4)$$

one introduces the coarse-graining integration with the characteristic function in the form of the Riesz fractional derivative[56]

$$\chi(x_j)P(x, y, t) \Rightarrow [\mathcal{W}_-^{1-\nu} + \mathcal{W}_+^{1-\nu}]P(x_j, y, t) = \mathcal{W}^{1-\nu}P(x_j, y, t). \qquad (B.5)$$

Appendix B.0.1. *Random fractal composite*

We consider a random fractal with an averaged volume, embedded in the $3D$, which is a function of a radius only $\mu(r) \sim r^{D_{\text{fr}}}$,[63,64] where $r = \sqrt{\sum_j x_j^2}$. Therefore, the distribution function and the kernel of the fractional integration are the radial functions, and thus, fractional integrations over the Cartesian coordinates x_j are substituted by integrations over the radial functions. This averaging procedure can be performed in the Fourier space as follows

$$\hat{\mathcal{F}}\left[\prod_j W^{1-\nu}P(r, y, t)\right] = \prod_j |k_j|^{1-\nu} \hat{P}(\{k_j\}, y, t) \Rightarrow k^{3-D_{\text{fr}}} \bar{P}(k, y, t), \quad (B.6)$$

where $k = \sqrt{\sum_j k_j^2}$ is the radius in the Fourier space and $\bar{P}(k, y, t) = \hat{\mathcal{F}}[P(r, y, t)]$. This averaging substitute in the $3D$ Fourier space is an extension of $1D$ embedding in the fractal, obtained in Refs. 62 and 48, in agreement with Nigmatulin's arguments on a link between fractal geometry and fractional integro-differentiation.[48,49] This is constituted in the procedure of averaging extensive physical values and expressed by means of a smooth function over a Cantor set that, eventually, leads to fractional integration.[48,49]

Note, that we did not use here any property of the kernel as a radial function that can be considered as the Riesz potential,[42] as well

$$\prod_j \partial_{|x_j|} \prod_j \frac{1}{|x_j - x_j'|^{1-\nu}} \Rightarrow \frac{1}{|x - x'|^{6-D_{\text{fr}}}} = \frac{\gamma(\alpha)}{\left(\sqrt{\sum_j (x_j - x_j')^2}\right)^{3-\alpha}}, \quad (B.7)$$

where $\alpha = D_{\text{fr}} - 3$ and $\gamma(\alpha) \equiv \gamma(D_{\text{fr}})$ is defined by Weber's integral

$$\int_0^\infty z^\beta J_\nu(z)dz = 2^\beta \Gamma\left(\frac{\nu + \beta + 1}{2}\right) \Big/ \Gamma\left(\frac{\nu - \beta + 1}{2}\right) \qquad (B.8)$$

at the Fourier transform of the Riesz kernel

$$\hat{\mathcal{F}}[r^{\alpha-3}] = \frac{(2\pi)^{\frac{3}{2}}}{\sqrt{k}} \int_0^\infty r^{\alpha-\frac{3}{2}} J_{\frac{1}{2}}(rk)dr = \gamma(D_{\text{fr}})k^{3-D_{\text{fr}}}. \qquad (B.9)$$

Following[42] (Ch. 25), we redefine $\prod_j W^{1-\nu}$ as the fractional degree of the Laplace operator $(-\Delta)^{-\alpha/2}$, namely

$$\prod_j W^{1-\nu} P(r, y, t) \Rightarrow \frac{1}{\gamma(\alpha)} \int \frac{P(r') \prod_j dx'_j}{|x - x'|^{3-\alpha}} \equiv (-\Delta)^{-\frac{\alpha}{2}} P(r, y, t). \quad \text{(B.10)}$$

This yields (see also Ref. 42)

$$\hat{\mathcal{F}}(-\Delta)^{-\frac{\alpha}{2}} P(r, y, t) = k^{-\alpha} \bar{P}(k, y, t). \quad \text{(B.11)}$$

Eventually, we arrive at the $3D$ comb model (26)

$$\partial_t P(r, y, t) = \delta(y) \Delta P(r, y, t) + [d\partial_y^2 - C](-\Delta)^{\frac{3-D_{\text{fr}}}{2}} P(r, y, t). \quad \text{(B.12)}$$

Appendix C. Fractional Integro–Differentiation

Fractional integration of the order of α is defined by the operator

$$_a I_x^\alpha f(x) = \frac{1}{\Gamma(\alpha)} \int_a^x f(y)(x - y)^{\alpha-1} dy, \quad \text{(C.1)}$$

where $\alpha > 0$, $x > a$ and $\Gamma(z)$ is the Gamma function. The fractional derivative is the inverse operator to $_a I_x^\alpha$ as $_a D_x^\alpha f(x) = {_a I_x^{-\alpha}}$ and $_a I_x^\alpha = {_a D_x^{-\alpha}}$. Its explicit form is

$$_a D_x^{-\alpha} = \frac{1}{\Gamma(-\alpha)} \int_a^x f(y)(x - y)^{-1-\alpha} dy. \quad \text{(C.2)}$$

For arbitrary $\alpha > 0$ this integral diverges, and as a result of a regularization procedure, there are two alternative definitions of $_a D_x^{-\alpha}$. For an integer n defined as $n - 1 < \alpha < n$, one obtains the Riemann-Liouville fractional derivative of the form

$$_a D_{RL}^\alpha f(x) = (d^n / x^n) {_a I_x^{n-\alpha}} f(x), \quad \text{(C.3)}$$

and fractional derivative in the Caputo form

$$_a D_C^\alpha f(x) = {_a I_x^{n-\alpha}} f^{(n)}(x). \quad \text{(C.4)}$$

When $a = -\infty$, the resulting Weyl derivative is

$$\mathcal{W}^\alpha \equiv {_{-\infty} D_W^\alpha} = {_{-\infty} D_{RL}^\alpha} = {_{-\infty} D_C^\alpha}. \quad \text{(C.5)}$$

One also has $_{-\infty} D_W^\alpha e^x = e^x$ this property is convenient for the Fourier transform

$$\hat{\mathcal{F}}[\mathcal{W}^\alpha f(x)] = (ik)^\alpha \hat{f}(k),$$

where $\hat{\mathcal{F}}[f(x)] = \hat{f}(k)$.

The Laplace transform can be obtained for Eq. (C.4). If $\hat{L}f(t) = \tilde{f}(s)$ is the Laplace transform of $f(t)$, then

$$\hat{L}[D_C^\alpha f(t)] = s^\alpha \tilde{f}(s) - \sum_{k=0}^{n-1} f^{(k)}(0^+) s^{\alpha-1-k}. \tag{C.6}$$

We also note that

$$D_{RL}^\alpha[1] = \frac{t^{-\alpha}}{\Gamma(1-\alpha)}, \quad D_C^\alpha[1] = 0. \tag{C.7}$$

The fractional derivative of a power function is

$$D_{RL}^\alpha t^\beta = \frac{t^{\beta-\alpha}\Gamma(\beta+1)}{\Gamma(\beta+1-\alpha)}, \tag{C.8}$$

where $\beta > -1$ and $\alpha > 0$. The fractional derivative from an exponential function can be simply calculated as well by virtue of the Mittag–Leffler function (see e.g. Ref. 41):

$$E_{\gamma,\delta}(z) = \sum_{k=0}^{\infty} \frac{z^k}{\Gamma(\gamma k + \delta)}. \tag{C.9}$$

Therefore, from Eqs. (B.9) and (C.9) we have the following expression

$$D_{RL}^\alpha e^{\lambda t} = t^\alpha E_{1,1-\alpha}(\lambda t). \tag{C.10}$$

References

1. H. Lodish, A. Berk, S. L. Zipursky, P. Matsudaira. D. Baltimore and J. Darnell, *Molecular Cell Biology* (W. H. Freeman and Company, New York, 2000).
2. AANS Classification of Brain Tumors, http://www.aans.org/.
3. D. Hanahan and R. A. Weinberg, *Cell* **100**, 57 (2000).
4. R. Stupp, W. P. Mason, M. J, van den Bent *et al.*, *N. Engl. J. Med.* **352**, 987 (2005).
5. R. Stupp, E. T. Wong, A. A. Kanner *et al.*, *Eur. J. Cancer* **48**, 2192 (2012).
6. A. Giese *et al.*, *Int. J. Cancer* **67**, 275 (1996).
7. A. Giese *et al.*, *J. Clin. Oncology* **21**, 1624 (2003).
8. A. Corcoran and R. F. Del Maestro, *Neurosurgery* **53**, 174 (2003).
9. T. Garay *et al.*, *Exper. Cell Res.* in Press (2013).
10. S. Khoshyomn, S. Lew, J. DeMattia, E. B. Singer and P. L. Penar, *J. Neuro-Oncology* **4**, 111 (1999).
11. A. Merzak, S. McCrea, S. Koocheckpour and G. J. Pilkington, *Br. J. Cancer* **70**, 199 (1994).
12. M. Tamaki *et al.*, *J. Neurosurg* **87**, 602 (1997).
13. L. Jerby, L. Wolf, C. Denkert, G. Y. Stein *et al.*, *Cancer Res.*, doi:10.1158/0008-5472. CAN-12-2215.
14. E. Khain and L. M. Sander, *Phys. Rev. Lett.* **96**, 188103 (2006).
15. H. Hatzikirou, D. Basanta, M. Simon, K. Schaller and A. Deutsch, *Math. Med. Biol.* **29**, 49 (2010).
16. A. Chauviere, L. Prziosi and H. Byrne, *Math. Med. Biol.* **27**, 255 (2010).

17. A. V. Kolobov, V. V. Gubernov and A. A. Polezhaev, *Math. Model. Nat. Phenom.* **6**, 27 (2011).
18. S. Fedotov, A. Iomin and L. Ryashko, *Phys. Rev. E* **84**, 061131 (2011).
19. E. Khain, L. D. Sander and A. M. Stein, *Complexity* **11**, 53 (2005).
20. C. A. Athale, Y. Mansury and T. S. Deisboeck, *J. Theor. Biol.* **233**, 469 (2005).
21. L. Zhang, Z. Wang, J. Sagotsky and T. S. Deisboeck, *J. Math. Biol.* **58**, 545 (2008).
22. L. Zhang, L. L. Chen and T. S. Deisboeck, *Math. Comp. Simulation* **79**, 2021 (2009).
23. A. Iomin, *Phys. Rev. E* **73**, 061918 (2006).
24. S. Fedotov and A. Iomin, *Phys. Rev. Lett.* **98**, 118101 (2007).
25. S. Fedotov and A. Iomin, *Phys. Rev. E* **77**, 031911 (2008).
26. M. Tektonidis *et al.*, *J. Theor. Biology* **287**, 131 (2011).
27. E. Khain *et al.*, *Phys. Rev. E* **83**, 031920 (2011).
28. E. D. Kirson *et al.*, *Cancer Res.* **64**, 3288 (2004).
29. E. D. Kirson *et al.*, *Proc. Nat. Acad. Sci. USA* **104**, 10152 (2007).
30. Y. Palti, *Europ. Oncological Disease* **1**(1), 89 (2007).
31. D. H. Geho *et al.*, *Physiology* **20**, 194 (2005).
32. A. Iomin, *Eur. Phys. J. E* **35**, 42 (2012).
33. A. Iomin, *Eur. Phys. J. Special Topics* **222**, 1873 (2013).
34. E. W. Montroll and M. F. Shlesinger, in J. Lebowitz and E. W. Montroll (eds.) *Studies in Statistical Mechanics*, V. 11 (North–Holland, Amsterdam, 1984).
35. R. Metzler and J. Klafter, *Phys. Rep.* **339**, 1 (2000).
36. A. Iomin, *J. Phys.: Conference Series* **7**, 57 (2005); *WSEAS Trans. Biol. Biomed.* **2**, 82 (2005).
37. E. W. Montroll and G. H. Weiss, *J. Math. Phys.* **6**, 167 (1965).
38. G. H. Weiss and S. Havlin, *Physica A* **134**, 474 (1986).
39. S. V. Petrovskii and B.-L. Li, *Exactly Solvable Models of Biological Invasion* (Chapman & Hall, Boca Raton, 2005).
40. J. D. Murray, *Mathematical Biology* (Springer, Heidelberg, 1993).
41. I. Podlubny, *Fractional Differential Equations* (Academic Press, San Diego, 1999).
42. S. G. Samko, A. A. Kilbas and O. I. Marichev, *Fractional Integrals and Derivatives* (Gordon and Breach, New York, 1993).
43. H. Bateman and A. Erdélyi, *Higher Transcendental Functions*, Vol. 3 (McGraw-Hill, New York, 1955).
44. S. Fedotov, A. O. Ivanov and A. Y. Zubarev, Non-homogeneous random walks and subdiffusive transport of cells, arXiv:1209.2851[cond-mat.stat-mech].
45. G. Zumofen, J. Klafter and A. Blumen, *Chem. Phys.* **146**, 433 (1990); G. Zumofen and J. Klafter, *Phys. Rev. E* **51**, 1818 (1995).
46. A. P. Prudnikov, Yu. A. Brychkov and O. I. Marichev, *Integrals and Series, Special Functions* (Gordon and Breach, New York, 1986).
47. I. M. Sokolov, J. Klafter and A. Blumen, *Phys. Today* **55**(11), 48 (2002).
48. R. R. Nigmatulin, *Theor. Math. Phys.* **90**, 245 (1992).
49. A. Le Mehaute, R. R. Nigmatullin and L. Nivanen, *Fleches du Temps et Geometric Fractale* (Hermes, Paris, 1998), Chap. 5.
50. J. H. P. Schulz, E. Barkai and R. Metzler, *Phys. Rev. Lett.* **110**, 020602 (2013).
51. E. Barkai, *Phys. Rev. Lett.* **90**, 104101 (2003).
52. A. Iomin, *Phys. Rev. E* **83**, 052106 (2011).
53. D. R. Cox and H. D. Miller, *The Theory of Stochastic Processes* (Methuen & Co. Ltd, London, 1970).
54. M. F. Shlesinger, *J. Stat. Phys.* **10**, 421 (1974).

55. A. Blumen, J. Klafter and G. Zumofen, in *Fractals in Physics*, eds. L. Pietronero and E. Tosatti (North–Holland, Amsterdam 1986), p. 399.
56. E. Baskin and A. Iomin, *Chaos, Solitons & Fractals* **44**, 335 (2011).
57. Yu. Polevaya, I. Ermolina, M. Schlesinger, B.-Z. Ginzburg and Yu. Feldman, *Biochimica et Biophysica Acta* **1419**, 257 (1999).
58. B. Rubinsky, G. Onik and P. Mikus, *Technol. Cancer Res. Treat.* **6**(1), 37 (2007).
59. A. K. Sarychev and V. M. Shalaev, *Electrodynamics of Metamaterials* (World Scientific, Singapore, 2007).
60. M. I. Stockman, *Physics Today* **64**(2), 39 (2011).
61. E. Baskin and A. Iomin, *Europhys. Lett.* **96**, 54001 (2011).
62. J. R. Liang, X. T. Wang and W. Y. Qiu, *Chaos, Solitons & Fractals* **16**, 107 (2003).
63. M. V. Berry and I. C. Percival, *Optica Acta* **33**, 577 (1986).
64. D. ben-Avraham and S. Havlin, *Diffusion and Reactions in Fractals and Disodered Systems* (University Press, Cambridge, 2000).

Chapter 5

Using Physics to Diagnose Cancer

Veronica J. James

Research School of Chemistry, Australian National University
Canberra 0200, Australia
veronica.james@anu.edu.au

This discussion about diagnostic tests for cancer incorporates a powerful branch of Physics namely X-ray diffraction. Although this technique was used to solve the DNA structure using the X-ray diffraction pictures of Rosalind Franklin,[1] and the structure of vitamin B12 by Dorothy Hodgkin[2] and hosts of other medical related structures, it is poorly understood by the general medical profession and the community at large. To the nonphysicist the patterns appear to have no relation to the results produced. It might as well be written in Greek. The well-known quote of Poincaré, the famous French mathematician and scientist, in 1885 comes to mind:

> "Science is built up with facts as a house is with stones.
> But a collection of facts is no more
> a science than a heap of stones is a house."

In order therefore to build a true understanding of this powerful technique it is necessary to build a firm understanding of the basic facts about this technique, so that the final results will be clear to all, as they will be held up by a firm house of knowledge. So let us take up the first stone.

First Stone: What is Diffraction?

Going back to base, therefore, we ask the question: "What is diffraction?" The answer in most textbooks is the bending of waves around gaps or obstacles in their paths. This term "diffraction" was introduced into Physics in our last years at high school but has long since been forgotten. Diffraction of water waves can easily be seen at break waters and at narrow entrances where we can also easily appreciate the much clearer effect for narrow openings.

However, in 1803 when Thomas Young discovered diffraction of light experimentally, light was considered to be a stream of particles as defined by Isaac Newton. Christiaan Huygens had proposed a wave theory for light in 1678 but this had been totally rejected. So when Young noticed that, if white light was passed through a narrow slit, bright bands of colour appeared in areas that should have been completely in shadow. This was similar to waves in the sea hitting walls at the entrance to a harbour and then being seen behind the walls as indicated in Fig. 1. If light did not bend, this was impossible.

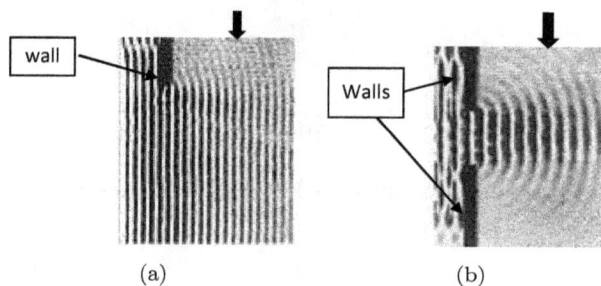

(a)   (b)

Fig. 1.   Diffraction of water waves.

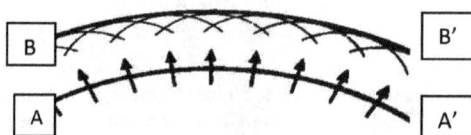

Fig. 2.   Wave nature of light proposed by Huygens.

This discovery was crucial in demonstrating the wave nature of light proposed by Huygens namely that every point on a wavefront may be considered as the source of a secondary wavelet. These secondary wavelets spread out in all directions at the speed of propagation of the original wave. The new wavefront is the envelope of these wavelets as demonstrated in Fig. 2. So applying the Huygens wave theory to a single slit, every point in the slit will become the source of a secondary wave.

Then followed the electromagnetic theory in the latter half of the 19th century and the controversy with the ether scientists. Finally the wave and particle theories were both accepted after Davison and Germer successfully tested De Broglie theory of a joint particle–wave nature for electromagnetic waves in 1926 by reflecting electrons from aluminium. The result was almost identical to that obtained by scattering X-rays from the aluminium lattice. This also confirmed Bragg's theory developed in 1913–1914 which opened up the X-ray diffraction studies to establish the crystal and molecular structures of small crystals.

Since the diffraction patterns accurately determine the positions of the molecules, only one true pattern can be obtained, the quality of the patterns obtained is the only difference that can exist.

Stone 2: Why X-Ray Diffraction?

For the best diffraction, the size of the gap or object should be roughly equal to the wavelength. This then acts as a solitary wave source with the wave spreading out completely in all forward directions. The spacing between carbon–carbon atoms or carbon–oxygen atoms is approximately 1.5 Angstroms (Å) or 0.15 nm so the obvious choice is a CuKα X-ray beam which has a wavelength of 1.5418 Å if the α and β beams are not resolved or 1.5405 Å if CuKα_1 is resolved and can be selectively chosen from the X-rays coming off a Cu anode. For all ordered crystals the atoms are arranged in sets of 3D planes.

Looking at the diagram below (Fig. 3), we can see that the beam bouncing off the top surface, at an angle α to the surface, has travelled a less distance than that bouncing of the first layer by a total distance of $2d \sin \alpha$. If this distance is 1 wavelength (λ) these waves will be in phase and produce a spot when focussed on a distant wall. This will also mean that 2 waves which are 2 planes apart will have a distance of 2 wavelengths and also be in phase and so on. Bragg's equation states that spots result when $a \sin \alpha = m\lambda$ where m is an integer $(1, 2, 3, \ldots)$ and the order number.

This was the area of my introduction to X-ray diffraction. Within two years it was extended to X-ray and neutron diffraction. Among the crystals solved using X-rays to locate the heavy atoms and neutrons to locate the hydrogen atoms was the crystal structure, trans-4-t-butylcyclohexyltosylate $C_{17}H_{26}SO_3$ (47 atoms). This was the largest crystal structure solved using the combination of X-ray and neutron diffraction at this time to produce residuals of not only 4.75% for X-rays but also 4.4% for neutrons. This accuracy for a neutron structure was the highest that had been achieved in 1969 and was highly praised by Professor G. E. Bacon, the pioneer of neutron diffraction, who had written the accepted textbook for Neutron Diffraction,[3] in 1955. Such simple crystal structures can now be solved almost completely by a computer using "the direct methods" programs. Using these programs I was involved in the solution of 40 structures.[4–43]

Scientists had also started looking at other samples, first powders in 1916 and then fibres in 1931. These fibres were naturally occurring such as muscle, hair, tendons, silk, cellulose, *etc.* or man-made such as actin filaments, microtubules, nerves, and collagen or synthetic polymers and DNA.

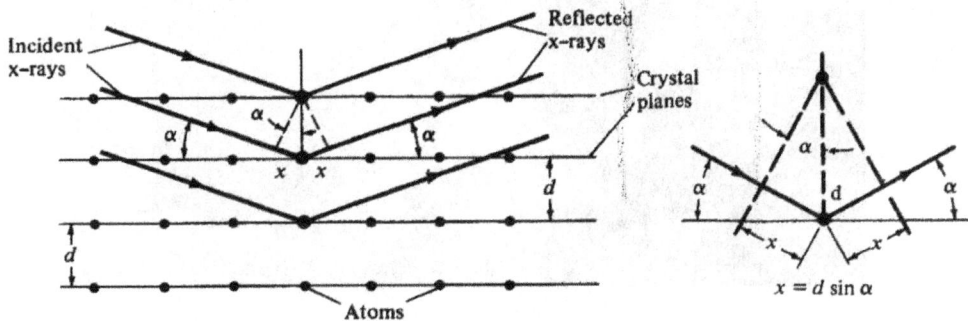

Fig. 3.   X-ray diffraction.

Stone 3: Diffraction of Collagen

Working with the late Dr. J. Halley using a polarising microscope to view pathology sections gave me convincing evidence that collagen in the breast changes in the lead up to breast cancer. These changes occur around the duct in which the breast cancer is located. It **did** stain for collagen and had been wrongly named elastotic tissue. There was no fat in this tissue since fat shows up under a polarising microscope. Just as a piece of meat that has no fat is very tough, this collagen is so dense that a hypodermic needle would bend rather than pass through it. As nature abhors a vacuum, if this tissue was near the surface, a dimple would appear on the surface when fat was removed. Dr. Halley always commented that the oncologist should look at the breast over the woman's shoulder. If such dense collagen appears in a frozen section, one should search for cancer.

Under the polarising pattern, the fat appears as yellow, the cancer appears as green, and stained collagen appears as red. These areas are clearly seen in Fig. 4.

Further evidence of the removal of fat and the consequential contraction of the tissue can be seen in the presence of two "elastotic" areas around large ducts on the one slide from a frozen section as indicated by arrows in Fig. 5. Two such ducts are not usually seen on one slide.

A 6 months sabbatical leave in 1978 at the European Molecular Biology Laboratory Grenoble, working on the effects of procaine, benzocaine and lignocaine on lecithin/cholesterol membranes with Andrew Miller and the late Carmen Berthet produced exciting results.[44,45] However these two scientists were experts in the study of collagen and in discussion with them about the changes in breast tissue

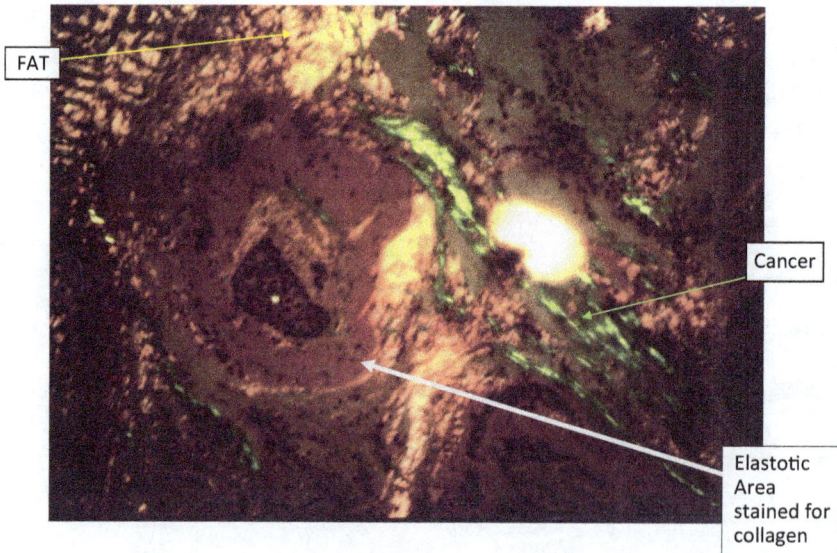

Fig. 4.   A polarising microscope slide from frozen sections.

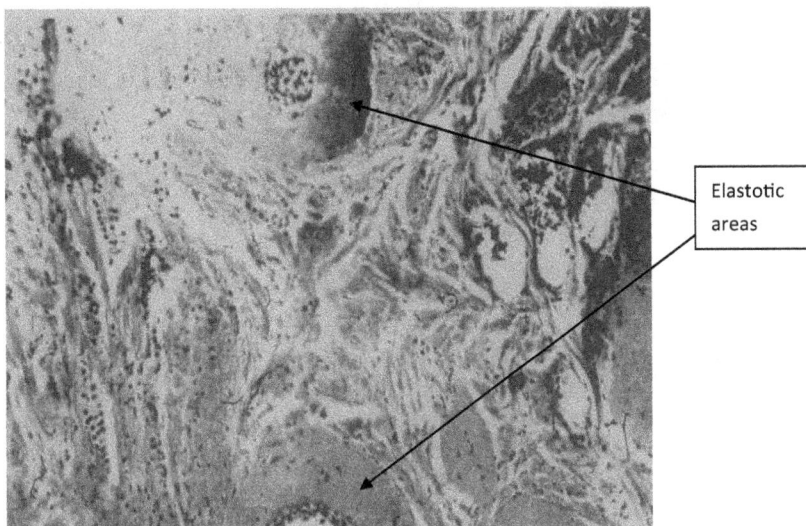

Fig. 5. The 2 elastotic areas surrounding large ducts are not usually seen on the same slide of a frozen section.

associated with breast cancer, I established the possibility of looking at these changes in the breast even though they thought that this would be very difficult. They helped me with a design of a cell for mounting collagen that maintained 100% humidity and allowed the sample to be stretched to remove the crimp. Such cells, shown in Fig. 6, have been used by me throughout all my research. Amongst other research they told me that they were involved in determining the structure of collagen Type 1 using rat tail tendon. The repeat gap-overlap in collagens is the repeat pattern which gives the D spacing for collagen, see Fig. 7.

They suggested that I should start with rat tail tendon as it is very ordered collagen Type 1. Rat tail tendon samples have been loaded first in all my research since and used to calculate the distance from sample to detector or film. The tendon collagen of rat tail is composed of fibrils which are slightly crimped. This crimp can be removed by stretching it slightly giving an almost crystalline pattern, see Fig. 8. One fibril is pulled out of the tail, placed in the cell and water added. This sample is ready for the experiment. Similar results were obtained with human and animal tendons where the collagen is arranged in parallel sheets which are slightly crimped.

The results obtained using these samples have included changes in tendon and skin with diabetes and ageing in humans and changes in foetal skin and tendon.[46] In addition to this study, a study was made on the very ordered chordae tendineae, the results of which showed the various stages of myxomatous heart valve degeneration. Surgeons had told me that the changes in myxomatous heart valves were the same as the elastotic change in breast cancer. In Fig. 9, (a) is of the diffraction pattern from a chordae tendineae taken from a normal heart valve. In (b), a mild myxomatous

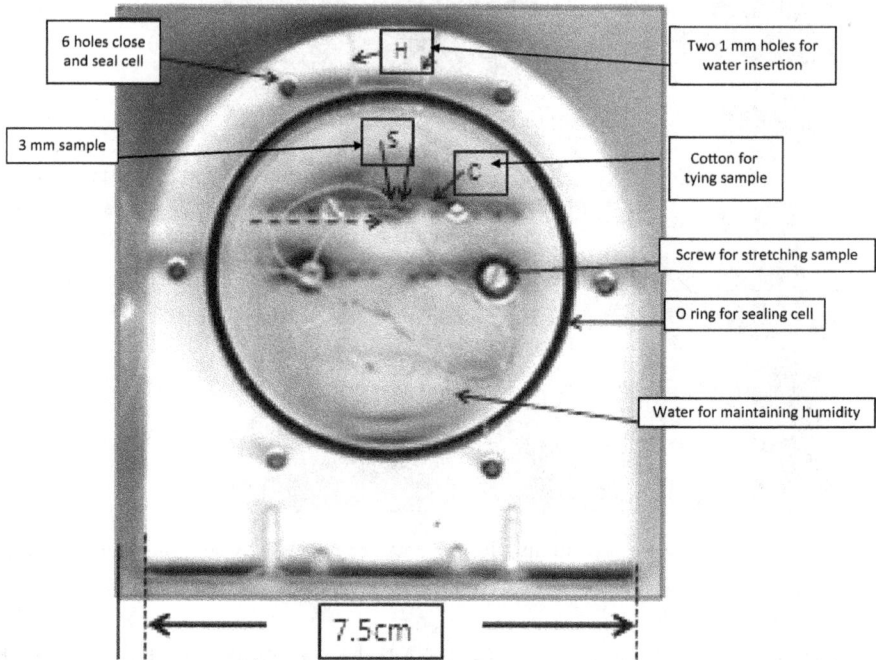

Fig. 6.   Cell for mounting collagen.

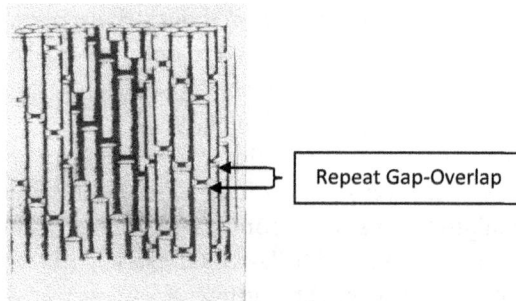

Fig. 7.   Structure of collagen Type 1 from rat tail tendon.

case, we note five additional peaks appear in the equatorial pattern, superimposed on the normal pattern, and indicated by arrow. Finally in (c), we have the case of a person who has reached the stage of valve collapse where peaks have now become circles. The meridional patterns were found to change only with age.[47]

A plot of intensities along the equatorial of Fig. 9(c) is given in Fig. 10, with intensities noted at peaks. An analysis of these results and those obtained from other samples gives possible periodicities of 438 Å or 762 Å. The first value here is the same as that of one of the rings found in breast tissue.

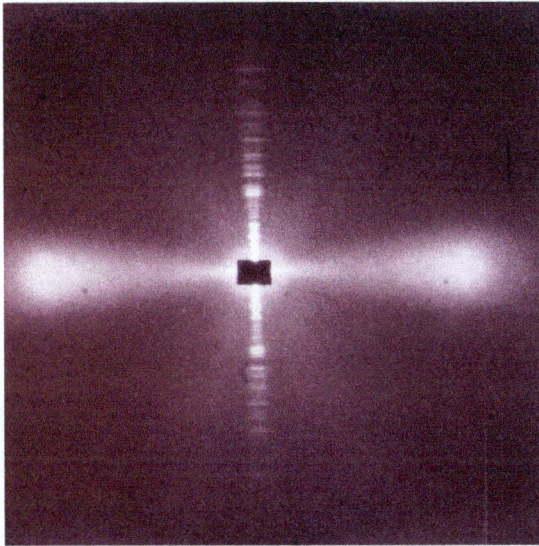

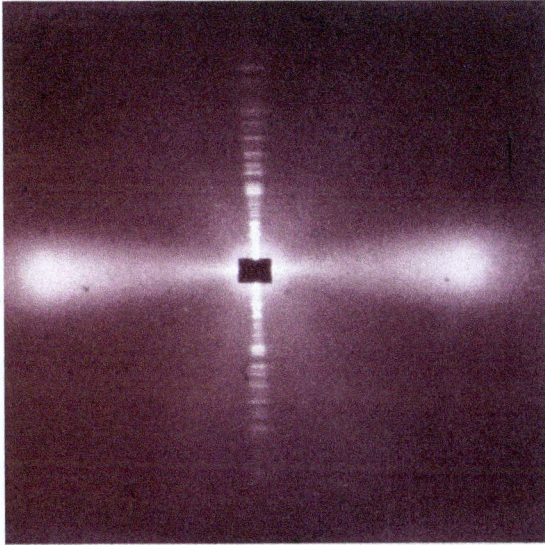

Fig. 8. Diffraction pattern of rat tail tendon.

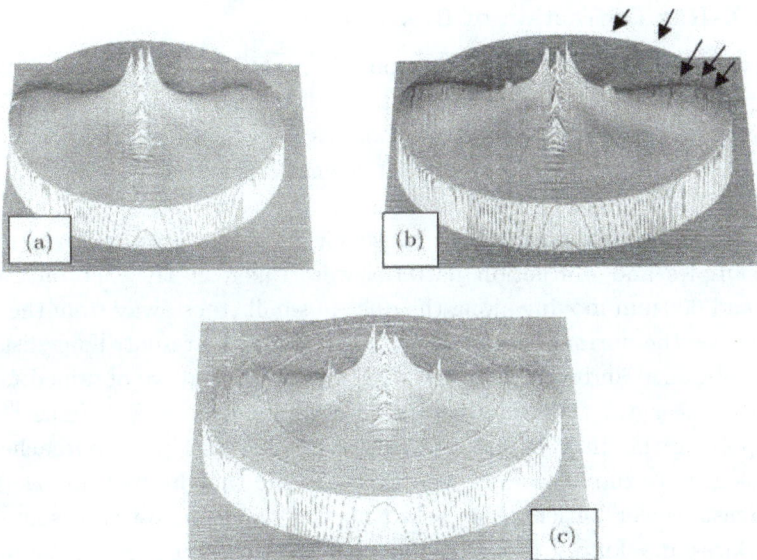

Fig. 9. Diffraction patterns of chordae tendineae: (a) A normal heart valve; (b) a mildly myxomatous heart valve; (c) heart of a myxomatous patient whose valve is about to collapse, the equatorial arcs are now rings.

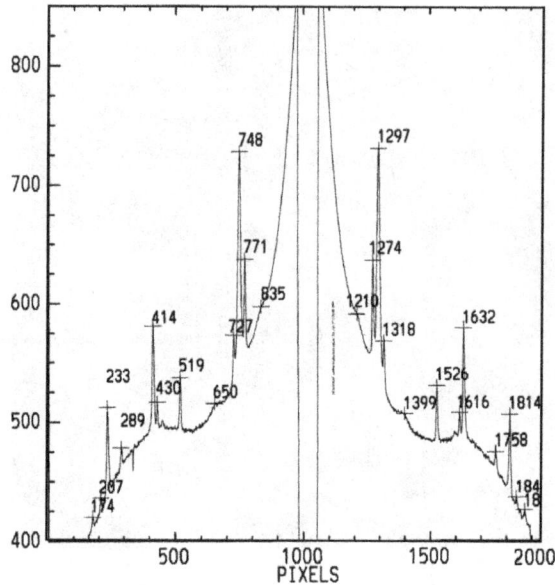

Fig. 10.   Plot of intensities along the equatorial of Fig. 9(c).

Stone 4: X-Ray Diffraction of Breast Tissue

Since the collagen fibres in Fig. 11 are roughly parallel to the duct, in order to get the best patterns, we need to select sections of ducts. These were obtained from pathology for mastectomy patients after ethics and consents. These sections were stretched to remove the crimp and the following diffraction pattern, Fig. 12, was obtained.

After ethics approval and patient consents, 120 ducts were removed from mastectomy samples and diffraction patterns were taken at the Photon Factory at 2400 mm and 600 mm moving along the duct in small steps away from the end near cancer towards the normal breast tissue, where fat was present. Four distinct patterns were obtained and were reported.[48] The pattern that we obtained for section adjacent to cancer was the same picture we obtained from foetal tissue.[49]

On reporting this in a lecture at Christie's Hospital, I was approached by the head oncologist, Anthony Howell, who informed me that he had discovered foetal skin in breast cancer patients. He asked me if I would investigate some samples for him. Daresbury kindly provided the time for this study. After completion of this study, Dr. Howell came to the synchrotron and I read out my results while he consulted his notes. At the end he said we should dance as I had correctly identified all positive samples. He then told me that I had one false positive, a lady who was in the high risk group and was asking for a double mastectomy. Because she was in her early 20's they were reluctant to oblige but now they must rethink. He agreed to collect more samples to verify my results.

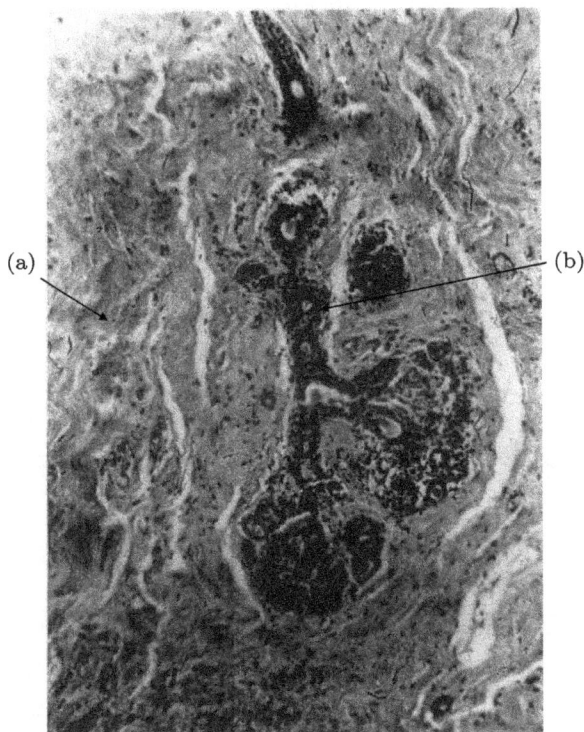

Fig. 11. A pathology slide of breast revealing that the collagen fibres (a) are roughly parallel to the duct (b).

Shortly after this I flew home, stopping in the USA and was horrified to find that my results being presented, without any mention of my name at a meeting, by a Daresbury computing expert, who had helped me set up the SAXS computing system to record my work. He had stolen my work. When I had gained some beam-time at the Photon Factory, I flew back to England to pick up the promised skin samples. The samples had disappeared from their freezer. I imagined they had been taken by the Daresbury computer person. I then asked Dr. Howell if he had his group of high risk people there. When he replied yes, I asked if he could collect some hair samples. Laughingly he asked what I hoped to see in hair. I replied, "Nothing, but I would lose face if I arrived for a specified beam-time with no samples". They were still laughing when I picked up the hair samples the following day. Arriving at the Photon Factory the next day, I was informed that a skin research paper had been published by the Daresbury team and had included one of my results from a Russian sample that I had taken at Tsukuba on my previous visit. I was not an author on that paper and, although I had been told that the samples supplied by Christie's hospital had to remain at Daresbury along with all computing results (not a normal procedure), I had removed my few Russian samples with me. They did not realize this was evidence of their stealth.

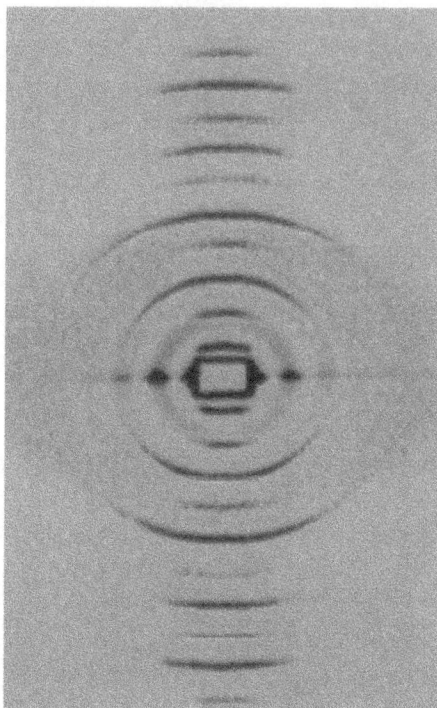

Fig. 12. Diffraction pattern of sections of duct from a mastectomy patient.

Stone 5: X-Ray Diffraction of Hair

When Professor Amemiya and I had set up the SAXS machine for my hair samples, he departed for Tokyo. I loaded the first sample and was amazed to find a ring superimposed on the normal hair pattern. Having worked on hair from diabetes patients[50,51] and Alzheimer's patients[52] where results did not show any rings, I thought such rings very strange. In diabetic patients, the glucose by-product was bound to the 626 Å helical sections, and in fact this change in the pattern helped identify that lattice, see Fig. 13. During the study of breast cancer, a different change appeared in the form of a set of spots fanning 7° in the equatorial pattern as seen in Fig. 14.[52] This change was shown to correlate with the presence of Alzheimer's disease appearing before any brain damage occurs.

I called Professor Amemiya and asked him what could be producing these strange rings. He responded, "You don't get rings in hair". He returned and after taking apart and remounting the diffractometer, another pattern was taken, followed by a second sample, with the same ring appearing. After repeating the procedure with a third sample, we finally had a pattern with no strange rings. Eight samples out of 20 produced those rings and I immediately sent an email to Dr. Howell with the numbers of these samples suggesting they might be from the same family or might use the same shampoo. He replied that these 8 were the

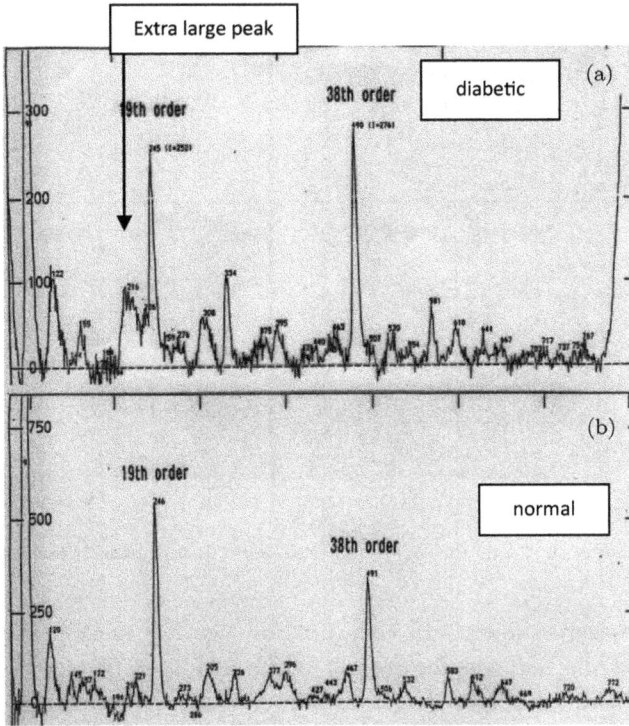

Fig. 13.   Plots of the intensities of the meridional patterns of hair for a normal patient and that of an insulin dependent patient with indicated meridional change.

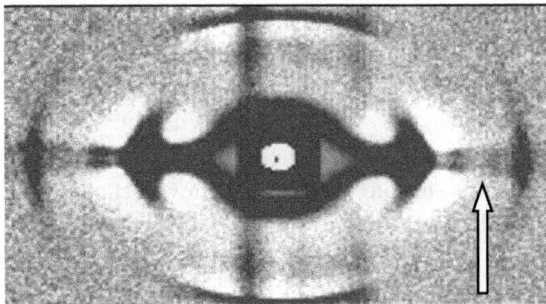

Fig. 14.   Diffraction patterns of hair for an Alzheimer disease patient, change indicated.

breast cancer patients. These results were published in Nature, March 4 1999.[53] Dr. Howell's name was removed from the paper by Nature because he had tried to interfere with its publication, so as to beat it into print with some results of his own. Subsequently, I was informed by Nature that he had sent a paper with patterns from 200 samples which he and Daresbury staff had tried to do.[54] The strongest 7th order was only present on a small number of the patterns, no other

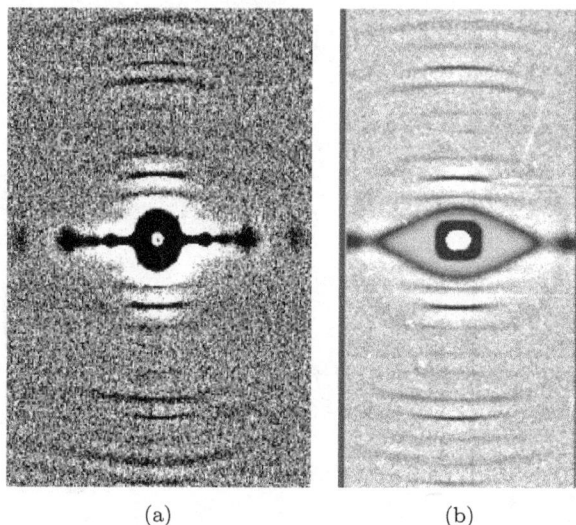

(a)   (b)

Fig. 15.   Diffraction patterns for (a) normal hair and (b) hair from breast cancer patient.

reflections were there. How could they possibly see the weak breast cancer ring? Fig. 15(a) shows the patterns for normal hair and Fig. 15(b) shows the weak ring superimposed on the normal pattern for hair from all breast cancer patients.

Subsequently I removed that ring by soaking the hair in Formic acid. A very thorough investigation with all available chemical techniques failed to show any extra material in the acid. One can only conclude that the material contents of the ring were broken into much smaller molecules, which remained inside the helices. Such molecules might possibly have given a ring which would be way outside the limits of a SAXS pattern.

The publication of the Nature paper immediately produced many papers from persons who had rushed to the synchrotrons with bundles of hair and had not achieved any results.[55–61] Only one of these groups[57] actually asked me to repeat the experiment using their samples. When I achieved the correct answers for these samples, they published a retraction of their paper.[58] To cap it off the following statement was written from a group of British scientists (Rogers, Hall, Hufton, Weiss, Pinder, Siu).

"*After reading and deliberating on the paper you recently published concerning breast cancer diagnosis by analysis of hair, we felt compelled to bring to your attention several of our concerns regarding this work and that presented elsewhere. We are troubled that to a nonspecialist, the correlation between the patient disease state and the diffraction features would seem high. This might lead to unjustified optimism over the rapid development of a diagnostic technique.*"

It is a great pity that none of these ever tried these experiments with me. Maybe they tried to do the experiment at Daresbury, where I could never get a pattern of hair. It seems wrong to me that none of these scientists ever asked me to show them how to do the experiment. With over 4000 samples now done and still no false negatives, I believe that the cruelty of mammograms should be a thing of the past along with corsetry.

So I stand accused of false results. I wrote to all the early groups and suggested what was wrong with their experiment and ways to fix the problem. I actually sat a day at APS, wrecking hair, back-combing, wetting and stretching it to indicate the difference between real rings and pseudo rings from poorly mounted hair and published the results.[62-64] I now understand how other scientists, *e.g.* Huygens, Max Boltzman, Max Feughelman, have been made to feel when they stepped outside the main road with new and correct but not accepted theories.

Stone 6: The Structure of Keratin

Working with Max Feughelman gave me great rewards as I had isolated the position of all meridional peaks in the hair diffraction pattern, see Fig. 16. These peaks obviously did not all belong to one lattice. Firstly I located those that belonged to the 470 Å lattice and put them in one column of an excel file. Finally I had put every peak into a lattice, see Table 1.

Working then with a colleague from the Physics school, Dr. Gleb Gribakin, a model was produced from which we calculated that for the known hexagonal

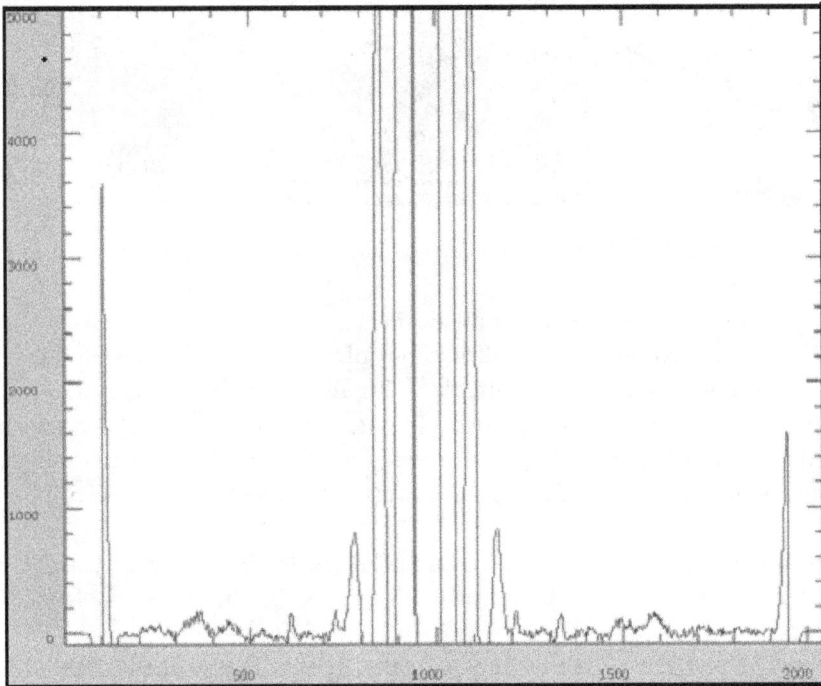

Fig. 16.   A plot of the intensities of the meridional pattern for hair.

Table 1.   Meridional lattice spacings.

D spacing	Origin of spacing	Observed orders
62.6 nm	Infinite lattice created by 3R cross-links	72
46.7 nm	Projected length of helical section of tetramer	50
19.8 nm	Projected length of 1/2 way point in helical section	20
27.2 nm	19.8 nm + stagger distance	26
12.2 nm	19.8 nm − stagger distance	10
7.8 nm	Stagger distance between ends of tetramers	buried
15.6 nm	Double stagger between ends of tetramers	buried

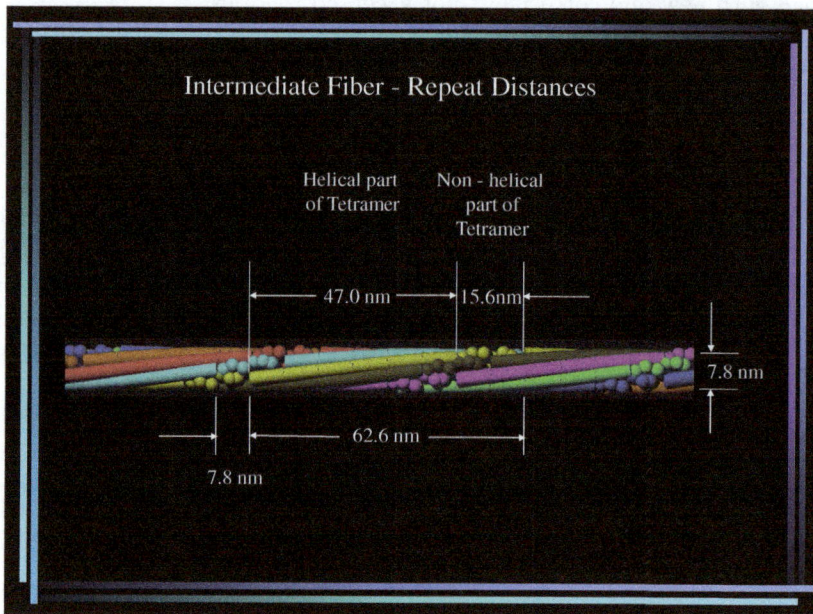

Fig. 17.   Illustrative diagram of different lattices in the hair.

structure the centre to centre distance apart of the fibrils must be 3 times the radius. This was proved for a small number of animals.[65] This was followed by a study of the hair or nails of 26 animals.[66] Finally Max Feughelman and I worked on a structure that agreed with both his data and mine.[67] This was checked by Dr. Barry Willis who built the confirmatory model and did the mathematics for Max Feughelman's findings. The mathematics of the model is given in full in a joint paper.[68]

This model was accepted by H. Zahn and R. B. D. Fraser but as yet not by any other scientists in the field. Maybe this will happen in 200 years time! It has been published.

Meanwhile a study of hair from colon cancer patients also showed an additional ring of slightly larger diameter. Study of colon sections revealed that changes were still visible in the ends of surgical sections which was different from that in the section near the tumour. It was suggested that a longer sample should have been removed to prevent a repeat of the tumour in this position in 2 years time. At this time such repeat tumours occurred in over 50% of cases and most were terminal when discovered. The surgeons subsequently sent another set of samples. All these were in fact normal at the ends. These results were published in full.[69]

Not only have my breast cancer diagnosis been correct[70] for all known cancer patients but some of the false positives have also been shown to be accurate. One such person had supplied samples in 3 successive years. The first two had shown false positives, in the third year it was correct but I was informed that she only had six months to live. She died that Christmas.

Some colleagues at Perth University were able to grow human breast cancers on the backs of nude mice. They removed whiskers before commencement of the implant and after blood was flowing through the implanted tumour. A diffraction study of these whiskers revealed no breast cancer changes until the blood started moving through the implant. This was convincing evidence that this change occurs earlier than any other test.[71]

This test can also verify the success or not of surgery. If the cancer is totally removed, the change will have gone as well. This had been noticed in the first set of samples examined for Christie's Hospital. One patient had undergone a mastectomy 7 years earlier and her operation was thought to be successful. Her sample did not show any change. Subsequently, 7 patients who had undergone mastectomies were followed for 7 years. The change was not found in 6 of them. The seventh one had 2 tumours at the beginning and pathology revealed a Type difference between the punch biopsy before the operation and that from mastectomy sample.

A further patient who had undergone a lumpectomy was strongly advised to undergo chemotherapy to follow. She did not wish to lose her waist-long hair but had it any way. The full length hair was supplied to me by her doctor. My study revealed that although the breast cancer ring was present before her lumpectomy, it was much fainter 8 days after and had disappeared completely 18 days after. The tragedy was that chemotherapy did not start until after that day. She need never have lost her beautiful hair.[72]

These changes in hair indicate that whatever the change which passes through the body must be, it must travel via the blood and be able to enter the hair at the time it is being formed in the follicle. Two reviews[73,74] have been written in the hope that someone might use this very accurate, highly successful method for the diagnosis of breast cancer. A venture capitalist company in Australia approached me in 2004 and offered to help me get this diagnostic test up and running. However they insisted that they must be given the patent. I agreed to do this on condition that it was noted in the agreement that I must be in charge of the science. They forgot that part of the agreement the moment that I signed it and appointed a man to that position who had only acted as a technician when he came to the synchrotrons with me. This man had had no experience in diffraction and only a B.Sc. in Analytical Chemistry but thought he could do this work as it looked easy.

He never succeeded and the company lost $30,000,000 when it went into bankruptcy in 2009. They refused to sell the patent to anyone who wanted to work with me but instead gave it to a company that is responsible for mammography.

In order to avoid this patent of mine, I investigated changes in nail clippings and found that similar changes to those already established with hair for breast and colon cancer and for Alzheimer's disease were present in the nails and took out a patent on the changes in the diffraction patterns of nails.

Tests of hair from a number of patients who had other cancers were carried out without success. Such cancers as melanoma, ovarian cancer and prostate cancer showed no change in hair at all. Since these three cancers have high incidence rates, I was disappointed. I wondered as breast cancer was also found in skin,[75] whether these three cancers might also show changes there as well. 3 mm skin biopsies were extracted from patients with these cancers after ethics approvals had been granted and consents obtained.

As yet no changes have been obtained for samples with ovarian cancer but changes that correlated specifically with melanoma and prostate cancer were discovered.[76,77] Melanomas have shown an extra ring located exactly on the 16th collagen order.

TRAMP mice, specifically bred to have prostate cancer, showed the prostate changes when 3 weeks old. Since prostate changes are not visible earlier than 10 weeks in any other test, this change should be seen at an earlier stage than is possible with any other techniques.[78]

Further tests have shown that there are specifically different and separate changes for BPH, for low and high grade prostate cancers, and for perineural and lymphatic invasions of prostate cancers.

These results have been submitted for publication and mark the end of my work to the present time.

Conclusion

The world awaits the introduction of these tests into the field of medicine. This technology will benefit everyone. Not only will mammograms be a thing of the past for women but PSA testing and biopsies of the prostrate will not be needed for men. Alzheimer's disease will be diagnosed early, before the brain is affected and therefore able to be treated. Colon cancer will be diagnosed early and removed completely, no reoccurrence. The world will be a much better place.

References

1. J. D. Watson and F. H. C. Crick, A structure for deoxyribose nucleic acid, *Nature* **171**, 737–738 (1953).
2. D. C. Hodgkin, J. Pickworth, J. H. Robertson *et al.*, The crystal structure of the hexacarboxylic acid derived from B_{12} and the molecular structure of the vitamin, *Nature* **176**(4477), 325–328 (1955).

3. G. E. Bacon, *Neutron Diffraction* (Clarendon Press, 1955).

4. V. J. James and F. H. Moore, A neutron diffraction study of trans-4-t-butylcyclohexyl-toluene-p-sulphonate $C_{17}H_{26}SO_3$, *Acta Cryst. B* **31**, 1053–1058 (1975).

5. K. Nimgirawath, V. James and J. D. Stevens, Methyl 1,2,3,4-tetra-0-acetyl-β-D-galactopyranuronate, $C_{15}O_{11}H_{20}$, *Cryst. Struct. Comm.* **4**, 617–622 (1975).

6. J. Ollis, V. J. James and J. D. Stevens, 1,2,3,4-tetra-0-acetyl-α-D-altopyranose, $C_{15}O_{11}H_{22}$, *Cryst. Struct. Comm.* **4**, 215–218 (1975).

7. J. Ollis and V. J. James, p-(di-2-chloroethyl) amino phenylbutric acid, chlorambucil., $C_{14}H_9Cl_2NO_2$, *Cryst. Struct. Comm.* **4**, 413–416 (1975).

8. V. J. James and K. Nimgirawath, Methyl-bis (β-chloroethyl) amine-N-oxide hydrochloride (nitromin), $C_5H_{12}C_{l3}NO$, *Cryst. Struct. Comm.* **4**, 41–44 (1975).

9. V. J. James and J. D. Stevens, 1,2,3,4-tetra-O-acetyl-β-D-arabinopyranose, *Cryst. Struct. Comm.* **3**, 19–22 (1974).

10. V. J. James and J. D. Stevens, *Cryst. Struct. Comm.* **3**, 187–190 (1974).

11. V. J. James and J. D. Stevens, Methyl 2,3,4-tri-O-acetyl-α-D-xylopyranoside, *Cryst. Struct. Comm.* **3**, 27–30 (1974).

12. V. J. James, Trans-3-t-butylcyclohexyl-toluene-p-sulphonate $C_{17}H_{26}SO_3$, *Cryst. Struct. Comm.* **2**, 205–208 (1973).

13. V. J. James and J. D. Stevens, 1,2,3,4-tetra-O-acetyl-8-D-ribofuranose, *Cryst. Struct. Comm.* **2**, 609–612 (1973).

14. V. J. James, The crystal and molecular structures of four pyranosides, in *Stockholm Symposium on the Structure of Biological Molecules*, Stockholm, Sweden, 9–11 July 1973.

15. V. J. James, t-t-decaly1-2-tosylate, $C_{17}H_{26}SO_3$, *Cryst. Struct. Comm.* **2**, 307–310 (1973).

16. P. L. Johnson, C. Cheer, J. P. Schaefer, V. J. James and F. H. Moore, The crystal and molecular structure of trans-4-t-butylcyclohexyl-toluene-p-sulphonate, *Tetrahedron* **28**(11), 2893–2899 (1972).

17. V. J. James and J. F. McConnell, The crystal and molecular structure of trans-4-t-butylcyclohexyl-toluene-p-sulphonate, *Tetrahedron* **28**(11), 2900–2906 (1972).

18. V. J. James and J. D. Stevens, An X-ray crystallographic study of methyl-0-Deribopyranoside, *Carbohydrate Research* **21**, 334–335 (1972).

19. V. J. James and C. T. Grainger, Cis-4-t-butylcyclohexyl-toluene-p-sulphonate $C_{17}H_{26}SO_3$, *Cryst. Struct. Comm.* **1**, 111–114 (1972).

20. V. J. James and J. F. McConnell, The crystal structure and conformation of cyclohexyltosylate, *Tetrahedron* **27**, 5475–5480 (1971).

21. V. J. James, F. H. Moore and J. D. Stevens, 1,2,3,4-Tetra-O-acetyl-β-D-arabinopyranose: A neutron refinement and comparison with X-ray refinement, *Australian Journal of Chemistry* **55**(2), 167–170 (2002).

22. C. J. Bailey, D. C. Craig, C. T. Grainger, V. J. James and J. D. Stevens, *Carbohydrate Research* **284**, 265–272 (1996).

23. S. J. Foster, V. J. James and J. D. Stevens, Methyl ß-D-glucoseptanoside, *Acta Cryst. C* **39**, 610–612 (1983); S. J. Foster, V. J. James and J. D. Stevens, Structures of 2 methyl B-D-glucoseptanoside derivatives, *Acta Cryst. C* **45**, 1329–1333 (1983).

24. R. A. Wood, V. J. James, J. D. Stevens and F. H. Moore, The crystal structure of 5-O-acety1-1,2,3:4,5-di-O-isopropylidene-α-Dglucoseptanose, $C_{14}O_7H_{22}$ by use of X-ray and neutron diffraction data, *Aust. J. Chem.* **36**, 2268–2277 (1983).

25. V. J. James and J. D. Stevens, Methyl 2,3,4,5-tetra-O-acetyl-ß-D-alloseptanoside, $C_{15}O_{10}H_{22}$, *Cryst. Struct. Comm.* **11**, 79–83 (1982).

26. V. J. James and J. D. Stevens, 1,2,3,4-tetra-O-acetyl-ß-DL-ribopyranose, $C_{13}O_9H_{18}$, *Cryst. Struct. Comm.* **11**, 457–460 (1982).

27. V. J. James and J. D. Stevens, Methyl 2,3,4-tri-O-acetyl-ß-D-lyxopyranoside, *Cryst. Struct. Comm.* **10**, 719–722 (1981).

28. V. J. James and J. D. Stevens, An X-ray crystallographic study of 5-O-acetyl-1,2:3,4 di-O-isopropylidene-α-D-galacto septanose, *Carbohydrate Research* **82**, 167–174 (1980).

29. D. C. Craig and V. J. James, 1,2,4,5/3,6-cyclohexanehexol (Muco-inisitol), $C_6O_6H_{12}$, *Cryst. Struct. Comm.* **8**, 629–634 (1979).

30. P. S. Clezy, D. C. Craig, V. J. James, J. F. McConnell and A. D. Rae, 10,10' Dith 10 bis (coproporphyrin II tetramethyl ester), $C_{80}H_{90}N_8O_{16}S_2$, *Cryst. Struct. Comm.* **8**, 605–608 (1979).

31. V. J. James, J. D. Stevens and F. H. Moore, Precision X-ray and neutron diffraction studies of methyl ß-D-ribopyranoside, $C_6O_5H_{12}$, *Acta Cryst.* B **34**, 188–193 1978.

32. J. Ollis, V. J. James and F. H. Moore, A neutron diffraction study of methyl 2,3:4,5-di-O-isopropylidene-α-D-glucoseptanoside, $C_{13}O_6H_{22}$, *Acta Cryst.* B **34**, (1978).

33. J. Ollis, V. J. James, S. J. Angyal and P. Pojer, An X-ray crystallographic study of α-D-allopyranosyl-α-D-allopyranoside, $CaCl_2 5H_2O$, a pentadentate complex, *Carbohydrate Res.* **60**, 219–228 (1978).

34. C. T. Grainger and V. J. James, Low angle X-ray studies of artificial biological membranes, in *Proceedings of the Australian Crystallographic Conference*, Bendigo, 1978.

35. V. J. James and J. D. Stevens, 1,2,3,4,6 penta-0-acetyl-β-D-gulopyranose $C_{16}O_{11}H_{22}$, *Cryst. Struct. Comm.* **6**, 119–122 (1977).

36. R. A. Wood, V. J. James and S. J. Angyal, The crystal structure of epi-inisitol strontium chloride complex, *Acta Cryst.* B **33**, 2248–2251 (1977).

37. V. J. James and J. D. Stevens, 1,2,3,4-tetra-0-acetyl-ct-D-ribopyranose, *Cryst. Struct. Comm.* **6**, 241–246 (1977).

38. V. J. James, F. H. Moore, J. Ollis and J. D. Stevens, Neutron diffraction studies on carbohydrates, *Proceedings of the Conference of Neutron Scattering, Gatlingburgh, Tennessee*, 1976.

39. V. J. James, K. Nimgirawath and J. D. Stevens, 1,2,3,4,6-penta-0-acetyl-β-D-xylopyranose, $C_{12}O_9H_{18}$, *Cryst. Struct. Comm.* **5**, 851 (1976).

40. J. Ollis, V. J. James, D. Ollis and M. D. Bogaard, Tetraphenyllarsonium tribromide $(C_6H_5)AS^+Br^-$. *Cryst. Struct. Comm.* **5**, 39–42 (1976).

41. K. Nimgirawath, V. J. James and J. A. Mills, Crystal structure of meso-L-glycero-L-gulo-heptitol, *J. Chem. Soc. Perkin Trans.* 2, **1976**, 349–35 (1976).

42. J. Ollis, M. Das, V. J. James, S. E. Livingstone and K. Nimgirawath, Tris (1,1,1-trifluoro-4-phenyl-4thiolo-but-3-en-2 ONE) cobalt(III),$C_{30}H_{18}F_9O_3S_3Co$, *Cryst. Struct. Comm.* **5**, 679–682 (1976).

43. R. A. Wood, V. J. James and J. A. Mills, 2,5,0-methylene-D-mannitol sodium chloride $C_{17}O_6O_{14}$ NaCl, *Cryst. Struct. Comm.* **5**, 207 (1976).

44. H. G. L. Coster, V. J. James, C. Berthet and A. Miller, Location and effect of procaine on lecithin/cholesterol membranes using X-ray diffraction methods, *Biochim. Biophys. Acta* **641**, 281–285 (1981).

45. H. G. L. Coster and V. J. James, Effects of local anaesthetics on lipid bilayer membranes, in *Proceedings of Small Angle Scattering International Conference* (Berlin, 2000), pp. 111–16.

46. V. J. James, L. Delbridge, S. V. McLennan and D. K. Yue, Use of X-ray diffraction in the study of human diabetic and ageing collagen, *Diabetes* **40**, 391–394 (1991).

47. V. J. James, J. F. McConnell and M. Capel, The D-spacing of collagen of mitral heart valves changes with ageing but not with collagen Type III content, *Biochim. Biophys. Acta* **1078**, 19–22 (1991).

48. V. J. James, Synchrotron fibre diffraction identifies and locates foetal collagenous breast tissue associated with breast carcinoma, *J. Synchrotron Radiation* **9**(2), 71–76 (2002).

49. V. J. James, J. F. McConnell and Y. Amemiya, Molecular structural changes in human fetal tissue during the early stages of embryogenesis, *Biochim. Biophys. Acta* **1379**, 282–288 (1998).

50. K. Wilk, V. James and Y. Amemiya, Intermediate filament structure of human hair, *Biochim. Biophys. Acta.* **1245**, 392–396 1995.

51. V. J. James, K. E. Wilk, J. F. McConnell, E. P. Baranov and Y. Amemiya, Intermediate filament structure of alpha-keratin in baboon hair, *Int. J. Biol. Macromolecules* **17**(2), 99–104 (1995).

52. V. J. James, J. C. Richardson, T. A. Robertson *et al.*, Fibre diffraction of hair can provide a screening test for Alzheimer's Disease: A human and animal model study, *Medical Science Monitor* **11**(2), CR53–57 (2005).

53. V. James, J. Kearsley, T. Irving, Y. Amemiya and D. Cookson, Using hair to screen for breast cancer, *Nature* **398**, 33–34 (1999).

54. A. Howell, J. G. Grossman, K. C. Cheung *et al.*, Can hair be used to screen for breast cancer? *J. Med. Genet.* **37**, 297–298 (2000).

55. F. Briki, B. Busson, B. Salicru, F. Esteve and J. Doucet, Breast cancer diagnosis using hair, *Nature* **400**, 236 (1999).

56. H. Amenitsch *et al.*, X-ray study at Trieste: No correlation between breast cancer and hair structure, *Synchrotron Radiat. News* **12**, 32–34 (1999) (unrefereed journal).

57. P. Meyer, R. Goergl, J. W. Botz and P. Fratzl, Breast cancer screening using small-angle X-ray scattering analysis of human hair, *J. Natl. Cancer Inst.* **92**(13), 1092–1093 (2000).

58. P. Meyer and V. James, Experimental confirmation of human hair scattering differences in breast cancer, *J. Natl. Cancer Inst.* **93**(11), 873–875 (2001).

59. B. Chu, D. Fang and B. S. Hsiao, Hair test results at the Advanced Polymer Beamline (X27C) at the NSLS, *Synchrotron Radiat. News* **12**, 36 (1999).

60. M. Capel, Hair test results at the NSLS, *Synchrotron. Radiat. News* **12**, (1999).

61. K. Laaziri, M. Sutton, P. Ghadirian *et al.*, Is there a correlation between the structure of hair and breast cancer or BRCA1/2 Mutations? *Phys. Med. Biol.* **47**, 1623–1632 (2002).

62. V. James, The importance of good images in using hair to screen for breast cancer, *J. Med. Genet.* **38**, e16 (2001).

63. V. J. James, Changes in the diffraction pattern of hair resulting from mechanical damage can occlude the changes that relate to breast cancer, *Phys. Med. Biol.* **48**, L37–L41 (2003).

64. V. J. James, The traps and pitfalls inherent in the correlation of the molecular structure of hair with breast cancer, *Phys. Med. Biol.* **48**, L5–L9 (2003).

65. V. J. James and Y. Amemiya, Intermediate filament packing in α-keratin of Echidna Quill, *J. Textile Technology* **68**(3), 167–170 (1998).

66. V. J. James, The molecular architecture for the intermediate filaments of hard α-keratin based on the superlattice data obtained from a study of mammals using synchrotron fibre diffraction, *Biochemistry Research International* **2011**, 198325 (2011).

67. M. Feughelman and V. J. James, Hexagonal packing of intermediate filaments in α-keratin fibres, *Textile Research Journal* **68**(2), 110–114 (1998).

68. M. Feughelman, B. Willis and V. James, in *Proc. 11th Int. Wool Research Conf.*, Leeds, UK, September 2006.

69. V. J. James, Fibre diffraction from a single hair can provide an early non-invasive test for colon cancer, *Medical Science Monitor* **9**(8), MT79–84 (2003).

70. V. James, False-positives in studies of changes in fibre diffraction of hair from patients with breast cancer may not be false, *J. N. C. I.* **95**(2), 170–171 (2003).

71. V. James, G. Corino, T. Robertson *et al.*, Early diagnosis of breast cancer by hair diffraction, *International Journal of Cancer* **114**(6), 969–972 (2005).

72. V. J. James, A review of low angle fibre diffraction in the diagnosis of disease, *British Journal of Medicine and Medical Research* **3**, 383–397 (2013).

73. V. J. James, Advances in understanding why the diffraction pattern of hair changes in breast cancer, *Expert Columnist Invitation, Expertos Columnas*, 6th December 2007.

74. V. J. James, A place for fibre diffraction in the detection of breast cancer? *Cancer Detection and Prevention* **2229**, 1–18 (2006).

75. V. J. James and B. E. Willis, Molecular changes in skin predict predisposition to breast cancer, *J. Med. Genet.* **39**(2), 1e (2002).

76. V. J. James and N. Kirby, The connection between the presence of melanoma and changes in fibre diffraction patterns, *Cancers* **2**, 1 (2010).

77. V. J. James, Fibre diffraction of skin and nails provides an accurate diagnostic test for malignancies, *International Journal of Cancer* **125**(1), 133–138 (2009).

78. V. J. James, Extremely early diagnostic test for prostate cancer, *Journal of Cancer Therapy* **2**, 77 (2011).

Chapter 6

From Single-Cell Dynamics to Scaling Laws in Oncology

Roberto Chignola* and Michela Sega†

Department of Biotechnology
University of Verona, Verona, Italy
*roberto.chignola@univr.it
†michela.sega@univr.it

Sabrina Stella

Department of Physics, University of Trieste
and INFN, Trieste, Italy
sabrina.stella@ts.infn.it

Vladislav Vyshemirsky

School of Mathematics and Statistics
University of Glasgow, UK
Vladislav.Vyshemirsky@glasgow.ac.uk

Edoardo Milotti

Department of Physics, University of Trieste
and INFN, Trieste, Italy
edoardo.milotti@ts.infn.it

We are developing a biophysical model of tumor biology. We follow a strictly quantitative approach where each step of model development is validated by comparing simulation outputs with experimental data. While this strategy may slow down our advancements, at the same time it provides an invaluable reward: we can trust simulation outputs and use the model to explore territories of cancer biology where current experimental techniques fail. Here, we review our multi-scale biophysical modeling approach and show how a description of cancer at the cellular level has led us to general laws obeyed by both *in vitro* and *in vivo* tumors.

1. The Life History of Tumors

It has already been pointed out that the time during which we study a tumor post-diagnosis is much shorter than the time from tumor initiation to diagnosis.[1] This important conclusion can be obtained from a straightforward analysis of clinical

---

*Present address: Department of Biotechnology, University of Verona, Strada le Grazie 15–CV1, I-37134 Verona, Italy.

data. It is a useful exercise because of its implications, and here we propose it again in a slightly different version and using more data.

The typical doubling time (DT) of human tumors is approximately 100 days (see Ref. 2), and if we accept as a first approximation that tumors grow exponentially, this means that the growth rate is $\log(2)/DT \simeq 6.9 \cdot 10^{-3}$ days^{-1}. If we assume a spherical cell shape and a typical cell radius of $8\,\mu\text{m}$ in human solid tumors,[3] the cell volume is $\simeq 2 \cdot 10^{-9}\,\text{cm}^3$. 90% of patients suffering from breast tumors die when tumors reach a volume of $\simeq 500\,\text{cm}^3$ (that is to say $\simeq 10\,\text{cm}$ diameter, see Ref. 4). On the other hand, it has been shown[5] that self- and clinical examination can detect breast tumors when their diameters are within 1.2 and 2.1 cm and thus, if we take on average a diameter of 2 cm, tumor volume at diagnosis is $\simeq 4\,\text{cm}^3$. Finally, a solid tumor can grow up to $\simeq 2\,\text{mm}$ in diameter in the absence of new blood vessels,[6] and hence at the end of the avascular phase its volume is $\simeq 4 \cdot 10^{-3}\,\text{cm}^3$. Thus, if a solid tumor grows up from a single cell we can estimate that it would take $\simeq 5.7$ years to end the avascular phase, $\simeq 2.7$ more years to become diagnosable and another $\simeq 1.9$ years to kill on average 90% of patients. The conclusion is that more than 80% of the life history of a solid tumor remains hidden.

During this time span tumor cells can evolve, new variants can be selected and migrate to distant sites to form colonies,[1] and in general solid tumors can acquire the aggressive phenotype that renders cancer a life-threatening disease. We can extend the limit of our knowledge by growing tumors in experimental animals, and even beyond that by growing three-dimensional aggregates of tumor cells *in vitro* to reproduce the biological properties of avascular tumors. However, in these cases it is not easy to take measurements at the appropriate spatial and temporal scales to investigate and understand the dynamics that govern the microscopic world of tumors.

When available, modern techniques force the experimenter to stop the growth process and destroy the cell aggregates to take measurements, and this leaves out the temporal dimension; on the contrary, if time is included in the assays and tumors are left free to grow then no measurements are possible to capture and quantify the fine details of the tumor microenvironment. Fortunately, there are tools to deal with this sort of biological uncertainty principle: present-day computers are powerful enough to tackle quite complex models, such as those that simulate tumors' behavior at multiple space and time scales.

2. Multi-Scale Cancer Modeling

The above considerations justify the efforts made by many scientists in their attempt to model tumor behavior: if we cannot observe tumors from appropriate time and space perspectives then we could try to model the life history of tumors from the very beginning and use the model to study the tumors' hidden properties. This simple yet powerful concept has fueled many modeling approaches, and we refer interested readers to Refs. 7–11 for comprehensive reviews.

One very basic and important assumption which is implicit in all growth models, or at least all the models that we are aware of, is that carcinogenesis has already taken place, so that each model describes cells that have already acquired the tumor phenotype: this is a powerful simplifying hypothesis in view of the complex series of events that lead to the emergence of the tumor cell phenotype, outlined, for instance, in two inspiring papers by Hanahan and Weinberg.[12,13]

Next, existing tumor models can be subdivided into three main classes depending on the underlying mathematics: continuous, discrete and hybrid continuous/discrete models. Continuous models, e.g. models based on differential equations, have many useful properties. They can be solved either analytically or numerically and their global properties can be investigated using methods from linear algebra and calculus to explore the entire phase space. Continuous models, by their very nature, necessarily deal with the average properties of the systems under study. The properties of biological systems, however, cannot always be averaged and the behavior of individual parts cannot always be neglected. For example, single mutants in a cancer cell population can acquire a new phenotype such as the ability to leave the primary tumor and colonize distant tissues, and this can determine the fate of the disease. In addition, life is often marked by important discrete events, for example, like cell division at mitosis, that are very difficult to model with continuous equations because of the large difference in the time scales of the processes involved. Finally, parameters in equations that model the global behavior of tumors often have no clear connection with the underlying biology, and hence they have no clear meaning. This often leads to continuous models with dimensionless parameters where no clear connection can be established between the vast mathematical phase space of the model and the narrow regions of biological interest.

Discrete numerical models can incorporate the behavior of individual cells. The main challenge in these models is the computational cost that may soon become too heavy in simulations that concern the growth from one cell up to 10^6 cells (the size of avascular tumors) or more than 10^{11} cells (the size of a clinically relevant tumor). To manage this cost the rules that govern the discrete numerical models are often overly simplified, they depart markedly from real biological processes, and risk falling into arbitrariness. As already pointed out by others,[11] some properties of human tissues — such as tissue biomechanics — are best modeled at the macroscopic scale using continuous models and, we add, even the bio-molecular pathways that determine the life cycle of the cells at the microscopic level are best modeled by the classical differential equations of chemical and enzyme kinetics. This is why hybrid discrete/continuous models have been developed. Biological systems are so complex because many different time and space scales are involved and because of the strong nonlinear interactions among variables at all scales. Thus each level may require different modeling approaches.

We believe that there is no golden standard in tumor modeling, so that eventually many different approaches might be required to develop useful models for oncology. In our own simulation program we decided to start with biophysical models at the cellular level. When validated with experimental data these models have been put together to obtain a quantitative *in silico* laboratory where we can now make interesting and potentially useful exploratory investigations.

3. A Virtual Biophysical Environment for Avascular Tumors

A detailed description of our multi-scale model of avascular tumors has already been given elsewhere.[14–17] We briefly recall here that our simulation program describes a realistic lattice-free environment where cells are free to move and eventually to exert both attractive and repulsive biomechanical forces on neighboring cells. The resulting motions of individual cells in the disordered tumor milieu can be followed in time while the overall tumor structure takes its shape. On the whole the growth dynamics of the simulated avascular tumors are determined by this collective behavior of cells.

At the same time, each cell lives, proliferates and dies. We have carefully modeled the complex biochemical networks that describe nutrient uptake and utilization by cells. Nutrients can either be converted to energy, or can be stored as energetic molecules in the cells, and along these pathways the nutrients are also converted into useful metabolites and waste products. Waste products are then secreted in the surrounding environment while both energy and metabolites are used to build up proteins, DNA and cellular structures. Among the synthesized proteins are cyclins and kinases that regulate the timing and the fate of the cell life cycle.[18,19] All these pathways have been carefully studied independently to fix model parameters and to reduce their known complexity to basic simplified reaction schemes. We have thus reduced the computational cost of the model and, at the same time, we have preserved its quantitative predictability. Thereafter each pathway has been connected to the others to obtain a metabolic model of the cell.[14,18,19] We proceed in a incremental way and include additional pathways as required to address specific aspects of tumor biology.

The basic actors of the model are nutrient and waste molecules which interact in the various intertwined biochemical pathways that regulate the cell's life. Each cell grows: the cell volume increases while the cell's materials (e.g. proteins, DNA, organelles) are built and the process is coordinated with the various phases of the cell cycle up to mitosis when the cell divides into two daughter cells. As in real cells, division is uneven and the mother cell's material is subdivided randomly between the two daughter cells.[14,18,19] This variability propagates to the whole cell population and determines the chaotic movements of the cells in the whole tumor and ultimately its final shape. All these clearly show that our model is really multi-scaled. Indeed, the model deals with objects that span at least 3 orders of magnitude in space (from a few μm of the cell radius up to a few mm of the diameter of an avascular tumor) and 12 orders of magnitude in time (from a few tens of μs for the characteristic times of the diffusion of molecular species up to

$\sim 10^7$s for the development of the tumor as a whole).[15] The computational problems that arise when one attempts to tame the numerical issues have been addressed elsewhere.[20]

Our computational approach stands out because of its quantitative aspects, and at present it involves approximately 100 parameter values. We have painstakingly searched the scientific literature to fix parameter values and when they were not available we estimated them by independent biophysical modeling of experimental data. Once fixed, the parameter values were not changed any more and simulation outputs were compared with new sets of experimental data to test the predictive behavior of the model and were successful. All comparisons between simulation outputs and experimental data have been carried out on a strict quantitative basis. We conclude that our simulation program is a reliable model of the growth of avascular tumors.

Interestingly, the search for parameter values has led us to take into consideration estimates obtained in very different experimental settings. For example, the parameters defining the activity of enzymes have often been measured in test tubes with crude tissue extracts; parameters defining the composition of the cell and its morphology have been measured in very different cell types. We finally realized that such a modeling effort can also be used to test the overall coherence of biological knowledge accumulated so far: as with a huge jigsaw puzzle we try to piece together biological information and check whether, at the end, it provides a coherent picture.

4. Using the Model to Investigate Tumor Biology

In this section we review some recent results that show how the model can be used to investigate the biology of tumors and to derive general laws. The multi-scale model can reproduce many aspects of tumor biology as far as metabolism (e.g. nutrient uptake and utilization, concentration gradients of nutrients and of waste molecules), proliferation kinetics and morphology are concerned at both the individual and at the population cell levels, and model outputs compare very favorably with actual experimental data.[14–19]

4.1. *A new tumor growth law*

At the cellular level, both *in vitro* avascular solid tumors and *in vivo* cancers display a layered structure.[21–24] The cells occupy different positions in the tumor tissue depending on several factors, but mainly dependent upon the chemical composition of the extracellular milieu. Actively proliferating cells accumulate in the tissue layers closer to the nutrient supply, no matter whether this is constituted by blood vessels or by a culture well filled with fresh medium, whereas quiescent cells distribute in the inner hypoxic and mildly acidic layers. The deepest territories are the necrotic areas that are mainly made up of dead cells and debris. This multi-layered structure, however, is not onion-like with layers separated by neat borders but is characterized instead by smoothly varying concentrations of alive cells in different phases of the cell cycle and of dead cells. These concentration changes depend only on the distance from the nutrient supply, and thus they unfold in the direction perpendicular to the supply interface.[21–24] Most importantly, these multi-layered structures determine relevant biological properties of solid tumors such as growth kinetics and sensitivity to therapeutic treatments.[23,24]

The multi-layered structure of solid tumors is well reproduced in our simulations, and since we can track the fate of each cell in the aggregate, we can also compute precisely the density of cells in a given phase at any depth. The results[25] show that the density $f(s)$ of live cells (i.e. the fraction of live cells per unit volume) decreases exponentially as we move further inside the cell cluster (Fig. 1).

$$f(s) = e^{-\frac{s}{\lambda}}. \tag{1}$$

A careful analysis shows that the parameter λ is not constant but is a weakly decreasing function of tumor size. However we have also shown that both the model with constant λ and with variable λ describe the growth of real avascular tumors just as well.[25]

What is important to note here is that cells have a certain probability to remain alive or die at a given distance s from the tumor surface according to the concentration of nutrients, such as glucose, amino acids, oxygen, and of waste products, such as lactic acid, in their surrounding environment. Cells reach this position dynamically as a result of the action of biomechanical forces (see above) that, in turn, are determined by cellular processes such as proliferation and division or death.

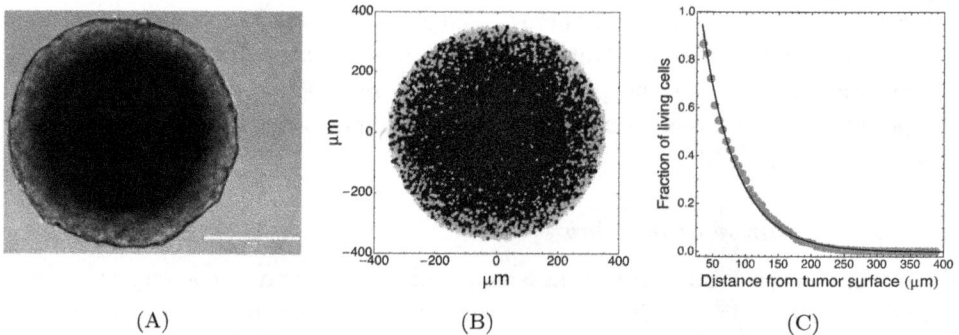

(A)   (B)   (C)

Fig. 1. Exploring the microenvironment of avascular tumors. Panel A: a micro-photograph of a three-dimensional avascular tumor grown *in vitro* with T47D cells (human breast carcinoma cell line). The white bar in the bottom-right corner is $400\,\mu m$ long. The inner darker area in the tumor is the necrotic area composed mainly of dead cells. Panel B: a simulated avascular tumor of approximately the same size as the one shown in Panel A. The figure shows the central section. Live cells are coloured in gray and dead cells in black. Panel C: fraction of living cells as the function of the distance from the tumor surface (symbols) observed for the simulated tumor shown in Panel B. The line shows the fit with Eq. (1).

Therefore, λ parameterizes the interactions between cells and their microenvironment as well as among cells and their neighbors.

From Eq. (1) above, we can obtain the expression for the total volume V_a occupied by live cells in a tumor of total volume $V = Ax^3$, where x is some characteristic length of the tumor (e.g. a chord that joins two recognizable, fixed features on the tumor surface) and A is the corresponding proportionality constant[25]:

$$V_a(x) \approx \int_0^x 3A(x-s)^2 e^{-\frac{(x-s)}{\lambda}} ds \approx \frac{3\lambda}{3\lambda + [V(x)/A]^{1/3}} V(x) = F(x)V(x) \quad (2)$$

where

$$F(x) = \frac{V_a(x)}{V(x)} = \frac{3\lambda}{3\lambda + [V(x)/A]^{1/3}} \quad (3)$$

is the fraction of live cells in the whole tumor which depends on tumor size at any given time.

The growth rate of tumor volume depends both on the proliferation rate α of live cells and on the known volume shrinking rate δ of dying cells. Therefore, taking into account the volume fractions occupied by live and dead cells, we find the growth law for tumor volume:

$$\frac{dV}{dt} = \alpha V_a - \delta V(1-F) = \alpha F V - \delta V(1-F). \quad (4)$$

The growth law given by Eq. (4) is straightforward, and yet it turns out to be a powerful descriptor of tumor growth when tested with experimental data[25] (see Fig. 2).

Indeed, a careful evaluation of Bayes factors showed that this model performs much better than the Gompertz growth model, a phenomenological descriptor of biological growth obtained by Benjamin Gompertz in 1825 (Ref. 26) and still a sort of "golden standard" among tumor growth models.[27-29] Most importantly, our model is based on parameters that have a clear biological meaning and whose values, therefore, provide important hints regarding the biology of solid tumors.

It is worth spending a few more words on the comparison between our model and the Gompertz model, since much work has been done in the past in an attempt to understand the biological foundations of the Gompertz model (see Ref. 29). The Gompertz model can be described by the following set of differential equations:

$$\begin{cases} \dfrac{dV}{dt} = \gamma_G(t)V(t) \\[2ex] \dfrac{d\gamma_G}{dt} = -\beta_G \gamma_G(t) \end{cases} \quad (5)$$

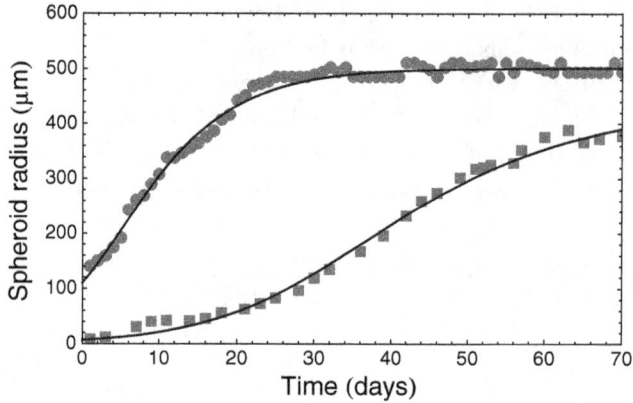

Fig. 2.   Time evolution of the radius of two avascular tumors obtained *in vitro* with 9L cells (rat glioblastoma cell line, circles) and MCF7 (human breast carcinoma cell line, squares). MCF7 tumors were obtained from cloned cells and thus, in this case, the initial radius corresponds to the radius of one cell. 9L tumors, on the other hand, were obtained with a different technique and they were grown for approximately one week before radii were finally measured. This is the reason why the initial radius in the 9L data set is higher than that of the MCF7 set. In both cases, solid lines are the result of a Bayesian regression of Eq. (4) obtained with the sequential Monte Carlo technique (details in Ref. 25).

where γ_G is the time-dependent growth rate and β_G is a constant. It turns out that our model can be recast into a similar system of differential equations[25]:

$$
\begin{cases}
\dfrac{dV}{dt} = \gamma(t)V(t) \\[2mm]
\dfrac{d\gamma}{dt} = -(\alpha + \delta)\left[\left(\dfrac{r(t)}{3\lambda} + 1\right)F(t) - 1\right]\gamma(t)
\end{cases}
\tag{6}
$$

where $r(t)$ is the tumor radius. The term in square brackets in Eq. (6) is actually quite close to a constant value over a wide range of r/λ values and the Gompertz model arises as an approximation of our growth law when we take $\beta_G \approx (\alpha + \delta)[(r(t)/3\lambda+1)F(t)-1] \approx$ constant. Thus, in the case of solid tumors, the Gompertz growth model naturally arises as an approximation of our biologically-motivated model.

4.2. *Metabolic scaling law of solid tumors*

Alive cells take up and consume nutrients. Since the total nutrient uptake μ of a tumor must be proportional to the number of live cells, and therefore to their total volume, we write:[30]

$$
\mu = \eta V_a = \eta F V
\tag{7}
$$

where η is the mean consumption rate per unit volume, and is given by $\eta = c/v_c$ where c is the mean consumption rate per cell and v_c is the mean cell volume.

Substitution of Eq. (3) into Eq. (7) yields:

$$\mu = \eta \frac{3\lambda V}{3\lambda + (V/A)^{1/3}}.$$ (8)

This equation shows that the nutrient consumption rate in solid tumors interpolates between a linear behavior $\mu \approx \eta V$ at small tumor size and a power law with exponent 2/3 at large tumor size: $\mu \approx 3\lambda\eta A^{1/3}V^{2/3}$.

Data on glucose uptake in different tumors are available and thus they can be used to check the validity of the scaling law given by Eq. (8). The mean consumption rate η may vary between different cell types because e.g. of the different expression of glucose transporters at the cell surface or because of the different energy demand. In addition, by fitting Eq. (8) to different data sets, we observed that parameter A varies greatly in different tumors while parameter λ is nearly fixed. These findings suggest a normalization of the metabolic rate to find a general scaling law for solid tumors[30]:

$$\hat{\mu}_N^{(k)} = \frac{\mu}{\eta_k A_k} = \frac{3\lambda(V/A_k)}{3\lambda + (V/A_k)^{1/3}} = \frac{3\lambda z}{3\lambda + z^{1/3}}$$ (9)

where η_k and A_k are the parameter values obtained for the kth data set and $z = V/A$. This law is obeyed by both *in vitro* avascular tumors and *in vivo* solid tumors of different types (see Fig. 3).

Surprisingly, this scaling law does apply to both avascular and vascularized tumors. The law depends on two free parameters, namely A and λ, if we exclude the specific metabolic rate of individual cells and their volumes. These quantities, however, can be easily measured with cell cultures and we did use experimental values in our calculations.[30] The parameter A determines both the total tumor volume

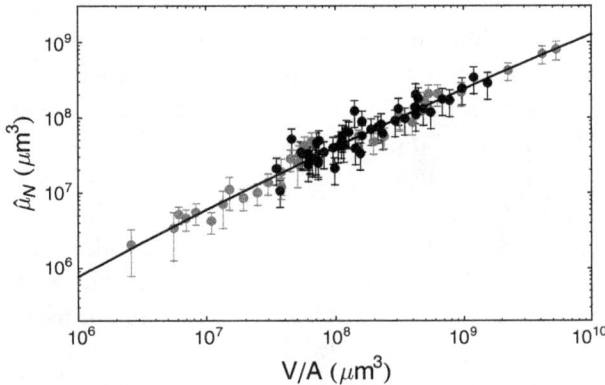

Fig. 3.   Normalized glucose consumption rate $\hat{\mu}_N$ vs. z ($z = V/A$, see Eq. (9)). Symbols refer to measurements of glucose uptake in avascular tumors *in vitro* (gray symbols) and in human tumors grown in immune-deficient rats (black symbols). Avascular tumors comprise spheroids obtained with different cell lines (details in Ref. 30). In this figure, however, we wish only to discriminate between avascular and vascularized tumors. The black line is a single fit with Eq. (9) to all data shown in the figure. The fit yields a common value $\lambda = 102 \pm 2$ μm for all tumors.

and the total tumor surface area, since $V = Ax^3$ and therefore $S = 3Ax^2$. The total tumor surface area corresponds to the boundary between the bulk of the tumor and the noncancerous environment, and this includes the interface between tumor and blood vessels even where they penetrate the tumor mass.[30] Thus, we expect higher values of A in vascularized tumors than in poorly vascularized cancers, as we did find in real data,[30] and thus a clear-cut classification between vascularized and avascular tumors based on parameter A. The normalized scaling law still contains the parameter λ and the same law, with fixed constant λ, appears to be obeyed by both avascular and vascularized tumors. As we have seen in the previous section, λ sets the spatial scale for the decrease of the fractional density of living cells. Moreover, the survival of cells at a given distance from the nutrient supply system does not depend on the supply itself, but rather on the chemical composition of the microenvironment that is heavily conditioned by the cells themselves. The results therefore seem to exclude the contribution of blood pressure or the permeability of tumor blood vessels as main determinants of the uptake and consumption of nutrients by tumors of a given size. We also remark that the model is not nutrient-specific and thus we expect the scaling law to apply to any molecule that is taken up and consumed by tumors, therapeutic drugs included.

5. Concluding Remarks

The multi-scale numerical model for avascular tumors has proved to be a useful tool to explore the dynamics of tumor microenvironments. Simulation experiments have led us to a new model of tumor growth and to a new metabolic scaling law which is also obeyed by vascularized tumors. The new scaling law may eventually be useful in clinical settings, as it shows how tumor volume, tumor growth kinetics and metabolism are related, and we note that glucose uptake and tumor size can actually be measured by imaging/radiological techniques. These results could not have been achieved without a strictly quantitative treatment of all the steps of model development. Such a prudent, quantitative approach provides a very direct comparison with experimental data, and leads to a highly reliable model.

The actual simulation runs are very time-consuming, but the efforts spent in taming the computational complexity of the multi-scale model have not been wasted, and we have simulated the growth of small avascular solid tumors up to a size of approximately 1 mm^3 (see Refs. 14–16). In practice, the algorithmic planning must be paralleled by the choice of a suitable programming framework. Since the very beginning of model development, we decided to write the simulation program in C++. This lets us take advantage of the many constructs of this object-oriented language in a way that naturally maps onto biological problems. Moreover, the extensive number of available C and C++ libraries is an added bonus that helps us to solve complex numerical problems such as the task of finding the proximity relations among cells. Indeed, since in our simulation program cells grow in three-dimensional space, it is necessary to regularly compute and update the reciprocal position between cells. This information is crucial for the calculation of the biomechanical forces among cells as well as for the processes that drive the diffusion of nutrients and other chemicals. The task is daunting, because we do not constrain our cells to regular lattice sites, and to this end we use the C++ library CGAL (www.cgal.org) that implements many sophisticated computational geometry algorithms. In particular, CGAL is used to calculate the Delaunay triangulation[31] that yields the proximity relationships. The triangulation algorithms implemented in CGAL are optimized and, in this way, we keep the computational complexity at bay: the time-complexity of the algorithm is only $O(N)$ rather than the worst case $O(N^2)$.

On the whole, computational complexity is a key issue, since the number N of cells grows nearly exponentially, and simulation time grows steadily as we simulate larger and larger tumors. For this reason, the computer program is presently under extensive revision and we hope, in the near future, to obtain a significant speedup as well as a much improved code modularity. While speedup is crucial to explore tumor evolution deep in time, the improved modularity will let us include many additional biological details into the model, and we shall be able to investigate different aspects of tumor biology and eventually the relationships between different cell types such as tumor and normal cells. A short list of studies that we plan to carry out after these improvements includes: (1) the simulation of the combined effects of anti-tumor drugs and radiotherapy to help plan better tumor control therapies; (2) the further inclusion of a basic model of tumor vascularization — like that developed in Ref. 32 — as a first step in the numerical description of the interaction of tumor cells with the surrounding tissues; (3) the extension of the present molecular network to some selected elements of the genetic regulation network: for instance, the phenomenological oxygen sensor in the present version of the program shall be replaced by the important network of the Hypoxia-inducible factors (HIF), which has been the focus of recent research for its potential relevance in whole-body protection against the adverse effects of radiation.[33]

Acknowledgments

We wish to acknowledge support from MIUR-PRIN2009 (Project: Numerical simulation of tumor spheroids), from the HPC CASPUR Standard Grant 2011, and from the University of Trieste, Italy — FRA 2013. Dr. Michela Sega and Dr. Sabrina Stella are the recipients of fellowships from MIUR (Ministero dell'Istruzione, dell'Università e della Ricerca).

References

1. J. E. Talmadge, *Cancer Res.* **67**, 11471 (2007).
2. E. Mehrara, E. Forssell-Aronsson, H. Ahlman and P. Bernhardt, *Cancer Res.* **67**, 3970 (2007).
3. J. P. Freyer and R. M. Sutherland, *Cancer Res.* **40**, 3956 (1980).

4. J. S. Michaelson, L. L. Chen, M. J. Silverstein, M. C. Mihm, A. J. Sober, K. K. Tanabe, B. L. Smith and J. Younger, *Cancer Res.* **115**, 5095 (2009).
5. U. Güth, D. J. Huang, M. Huber, A. Schötzau, D. Wruk, W. Holzgreve, E. Wight and R. Zanetti-Dällembach, *Cancer Epidemiol* **32**, 224 (2008).
6. D. Ribatti, A. Vacca and F. Dammacco, *Neoplasia* **1**, 293 (1999).
7. H. Byrne and D. Drasdo, *J. Math. Biol.* **58**, 657 (2009).
8. P. Tracqui, *Rep. Prog. Phys.* **72**, 056701 (2009).
9. H. Byrne, *Nature Rev. Cancer* **10**, 221 (2010).
10. K. A. Rejniak and A. R. A. Anderson, *Rev. Syst. Biol. Med.* **3**, 115 (2011).
11. T. S. Deisboeck, Z. Wang, P. Macklin and V. Cristini, *Annu. Rev. Biomed. Eng.* **13**, 127 (2011).
12. D. Hanahan and R. A. Weinberg, *Cell* **100**, 57 (2000).
13. D. Hanahan and R. A. Weinberg, *Cell* **144**, 646 (2011).
14. E. Milotti and R. Chignola, *PLoS ONE* **5**, e13942 (2010).
15. R. Chignola, A. Del Fabbro, M. Farina and E. Milotti, *J. Bioinf. Comput. Biol.* **4**, 559 (2011).
16. R. Chignola and E. Milotti, *AIP Adv.* **2**, 011204 (2012).
17. E. Milotti, V. Vyshemirsky, M. Sega, S. Stella, F. Dogo and R. Chignola, *IEEE/ACM Trans. Comput. Biol. Bioinf.* **10**, 805 (2013).
18. R. Chignola and E. Milotti, *Phys. Biol.* **2**, 8 (2005).
19. R. Chignola, A. Del Fabbro, C. Dalla Pellegrina and E. Milotti, *Phys. Biol.* **4**, 114 (2007).
20. E. Milotti, A. Del Fabbro and R. Chignola, *Comput. Phys. Commun.* **180**, 2166 (2009).
21. I. Tannock, *Br. J. Cancer* **22**, 258 (1968).
22. J. Moore, P. Haleton and C. Buckley, *Br. J. Cancer* **51**, 407 (1985).
23. R. Sutherland, *Science* **240**, 177 (1988).
24. W. Müller-Klieser, *Crit. Rev. Oncol. Hematol.* **36**, 123 (2000).
25. E. Milotti, V. Vyshemirsky, M. Sega and R. Chignola, *Sci. Rep.* **2**, 990 (2012).
26. B. Gompertz, *Phil. Trans. Royal Soc. London* **115**, 513 (1825).
27. A. Laird, *Br. J. Cancer* **18**, 490 (1964).
28. A. Laird, *Br. J. Cancer* **19**, 278 (1965).
29. Ž. Bajzer and V. Pavlović, *Comput. Math. Methods Med.* **2**, 307 (2000).
30. E. Milotti, V. Vyshemirsky, M. Sega, S. Stella and R. Chignola, *Sci. Rep.* **3**, 1938 (2013).
31. M. De Berg, M. Van Kreveld, M. Overmars and O. Schwarzkopf, *Computational Geometry: Algorithms and Applications* (Springer-Verlag, New York, 2000).
32. M. Welter and H. Rieger, *Eur. Phys. J. E* **33**, 149 (2010).
33. C. M. Taniguchi *et al.*, *Sci. Transl. Med.* **6**, 236ra64 (2014).

Chapter 7

Adhesion-Mediated Signalling in Cancer: Recent Advances and Mathematical Modelling

Olivia Crociani[*], Andrea Becchetti[†], Duccio Fanelli[‡]
and Annarosa Arcangeli[*,§]

[*]Department of Experimental and Clinical Medicine
Section of Internal Medicine, University of Florence
Centro Interdipartimentale per lo Studio di Dinamiche Complesse (CSDC)
Florence, Italy

[†]Department of Biotechnology and Biosciences
University of Milano Bicocca, Milano, Italy

[‡]Department of Physics and Astronomy
University of Florence, Florence, Italy
[§]annarosa.arcangeli@unifi.it

Cancer can be viewed as a "tissue", where neoplastic cells are immersed into a peculiar microenvironment (the "tumor microenvironment", TME) which modulates tumor cell behaviour during multistep tumorigenesis. Based on this concept, antineoplastic therapy should be tuned to target not only tumor cells but also the cellular constituents of the TME. Such necessity is well exemplified by considering tumor angiogenesis, a major aspect of cancer biology.

Ion channels and transporters are increasingly recognized as relevant players in the tumor cell-TME cross-talk. For example, during tumor neo-angiogenesis, soluble factors as well as fixed components of the extracellular matrix (ECM) and membrane proteins determine signal exchange between the TME and the implicated cell types. The signalling network is coordinated by functional "hubs", which may be constituted by integrin receptors associated with other proteins to form macromolecular signalling platforms at the adhesive sites. These complexes often include ion channels.

The K^+ channels encoded by the human ether-à-go-go related gene (Kv11.1, or hERG1) are frequently overexpressed in human cancers and regulate intracellular signalling by physically associating with integrin subunits and growth factor/chemokine receptors. In colorectal cancer (CRC) we recently identified a novel signalling pathway centered on hERG1 channels and integrins. This pathway involves the p53 protein, which is encoded by a tumor suppressor gene often mutated in human cancers. p53 controls angiogenesis, through a mechanism regulated by hERG1 K^+ channels. The central role played by hERG1 in CRC angiogenesis suggests that targeting hERG1 may be an effective therapeutic option in patients with advanced CRC.

To better understand the above process, it is necessary to study the interlaced dynamics of the key microscopic actors by using dedicated mathematical models. We here review a simple model, of reductionist inspiration, that explores the intimate connections between apoptosis and hypoxia, passing through angiogenesis. We show that a dynamical switch takes place between the normoxia and cellular death conditions. When oxygen

---

[§]Corresponding author.

lacks, cells can cross the transition line and so gain their way towards the normoxia regime, by implementing point mutations that affect the p53 production and activation rate, with the involvement of K^+ ion homeostasis, in agreement with the experimental observations.

1. Introduction

The traditional view of cancer as a collection of proliferating cells must be reconsidered. Cancer should be viewed as a "tissue", constituted by both transformed cells and a heterogeneous microenvironment, which is constructed and remodelled by tumor cells during the course of multistep tumorigenesis. This "tumor microenvironment" (TME) is a complex array of cells and extracellular matrix (ECM) proteins[1] that strongly affects the behaviour and malignancy of the transformed cells. Moreover, the TME may change during tumor progression; hence it may differ (structurally and functionally) between the primary tumor and its metastases.[2,3] The TME greatly varies among cancers of different histogenesis. For example, in leukemias, it is mainly represented by the bone marrow, with the complex array of stromal and vascular cells that constitutes the bone marrow niche, where leukemia stem cells reside.[4] In carcinomas, a clear distinction is made between the neoplastic cells (the "parenchyma") and the TME, indicated as the "tumor stroma". An active and overwhelming tumor stroma (causing the so-called "desmoplastic reaction") characterizes some specific carcinomas, such as breast, prostate or pancreatic cancer.[5]

Taking into account the relevance of the tumor stroma, antineoplastic therapy must be tuned to target the "cancer tissue", e.g. not only tumor cells but also the cellular constituents of the TME.[6,7] Such necessity is well exemplified by considering tumor angiogenesis, a major aspect of cancer biology. To ensure blood delivery to the cells located deep into the tumor mass, cancer cells promote angiogenesis since the early stages of neoplastic progression (the so-called "angiogenic switch").[8] Sustained angiogenesis also fosters the metastasis growth.[9] To this aim, tumor cells secrete pro-angiogenic factors. A major example is the Vascular Endothelial Growth Factor-A (VEGF-A; reviewed in Ref. 10). Transcription of the *VEGF-A* gene depends on the O_2 levels.[11] In particular, hypoxic conditions activate the Hypoxia Inducible Factors (HIFs; [12–14]). Under hypoxia, the HIF-α subunits (either HIF-1 or HIF-2) are protected by the von-Hippel Lindau (VHL)-dependent degradation in the proteasome.[15] In cancer, this pathway can be abnormally stimulated under normoxic conditions.[12] The process is known as hypoxic mimicry[16] and may depend on VHL mutations[17] or activation of oncogenes, such as the membrane receptors that regulate the PI3K/Akt pathway.[18,19]

In this context, one should also consider that ion channels and transporters (ICTs) are increasingly recognized as relevant players in the tumor cell-TME cross-talk.[20] Work carried out in the past three decades indicates that changes in expression and activity of a variety of ion channels are implicated in specific stages of the neoplastic progression. These latter range from altered cell proliferation to apoptosis and invasiveness. Particularly ample evidence is available for K^+ channels, whose expression is often altered in primary human cancers. These ion channels generally exert pleiotropic effects on the cell cycle machinery by modulating, for example, the Ca^{2+} fluxes and cell volume. However, ion channels are also involved in later tumor stages, through the regulation of angiogenesis, the cell-ECM interaction and cell motility. Many cellular mechanisms contribute to these physiological effects and ion channels are emerging as relevant players in the tumor cell-TME cross talk. In particular, ion channels can form protein complexes with other membrane proteins such as integrins or growth factor receptors, thus activating intracellular signalling cascades. In general, ion channels are placed in a pivotal position to sense and transmit extracellular signals into the intracellular compartment, thus cooperating with ECM receptors. Similar roles appear to be exerted by ion transporters, which contribute to regulate two features of the neoplastic tissue, i.e. hypoxia and acidic extracellular pH.

From a clinical standpoint, targeting of ion channels may be revealed to be an effective novel way to control the malignant progression, as recent evidence indicates that blocking channel activity impairs the growth of some tumours, both *in vitro* and *in vivo*.[21–23] In addition, altered channel expression can be exploited for diagnostic purposes or to convey traceable or cytotoxic compounds to specific neoplastic cell types.

2. Integrin Receptors and Ion Channels in Cancer

Integrins are membrane heterodimeric proteins formed by α and β subunits. Eighteen α and 8β subunits are currently known in mammals, which can form at least 24 heterodimers. Integrin subunits are type I glycoproteins comprising a 20–70 aminoacid cytoplasmic tail (1000 aminoacids for $\beta4$ [Ref. 24]), a membrane-spanning helix and a large multidomain extracellular portion.[25] Each heterodimer binds a specific ensemble of ECM proteins.

Besides permitting cell adhesion to the substrate, integrins transmit bidirectional signals across the plasma membrane. In resting conditions, integrins are in a low affinity state, and can be activated by "inside-out" signalling usually mediated by the intracellular proteins talin and kindlin. Talin binding to the β integrin cytoplasmic tail is thought to be the final step of integrin activation.[26] Conversely, integrin binding to its extracellular ligands produces the so-called "outside-in" signalling, which regulates cell motility, proliferation, differentiation, etc.[27] The literature on integrin-mediated signalling is immense and we will merely mention that integrins appear to be related to most of the known intracellular pathways, such as induction of cytosolic kinases, stimulation of the phosphoinositide metabolism, activation of Ras/MAPK and PKC pathways and regulation of Small GTPases.[25,28,29] Therefore, integrin signalling often overlaps with the effects exerted by growth factor or cytokine receptors.[29] This interaction allows cells to properly integrate the stimuli provided by the ECM, growth factors, hormones and mechanical stress, to produce physiologically meaningful responses. In cancer tissue, such integrative mechanisms determine the fate of tumor cells.

2.1. *Integrin relationships with ion channels*

The earliest indications of a direct interaction between integrins and ion channels came from studies on leukemic cells and neuroblastomas, in which adhesion-dependent differentiation or neurite extension were found to depend on activation of K^+ channels.[30–34] When associated with integrins, ion channel function becomes itself bidirectional, being regulated by extracellular signals (through integrins) and in turn controlling integrin activation and/or expression.[30] A similar pattern has been observed in some transporters mediating proton fluxes,[35–37] which establish the reversed H^+ gradient that characterizes neoplasias.[38] The integrin-channel cross talk may depend on diffusible cytosolic messengers commuting between the two proteins (reviewed in Ref. 30). A classic example is T lymphocyte activation, where $\beta 1$ integrins control Ca^{2+} influx, which at the same time regulates and is regulated by K^+ channel activity.[39] On the other hand, integrins trigger mechanical tension at focal adhesion sites by regulating not only Ca^{2+} signalling,[40] but also molecules such as FAK and c-Src.[41]

Subsequent work indicated that integrin receptors and ion channels can also physically associate to form supramolecular membrane complexes, which constitute signalling hubs that coordinate downstream cellular signals. The first evidence was obtained in T lymphocytes by Levite and colleagues,[42] who observed that $\beta 1$ integrins associate with Kv1.3 channels. A physical link between Kv1.3 and $\beta 1$ integrins was also described in melanoma cells.[43] In addition, we found that $\beta 1$ integrins can also associate with Kv11.1 (hERG1) channels, on the plasma membrane of either leukemias or solid cancers.[44–48] The molecular complex also involves growth factor/chemokine receptors and can recruit cytosolic signalling proteins, which thereby regulate cellular signalling. The complex formation has an impact on leukemia progression since it can either trigger chemoresistance[47] or control angiogenesis.[48]

Another mechanism involving the interaction between integrins and ion channels contributes to determine integrin recycling.[49] In particular, CLIC3 chloride channels colocalize with active $\alpha 5\beta 1$ integrins in late endosomes/lysosomes, allowing the integrin to be retrogradely transported and recycled to the plasma membrane at the cell rear. This mechanisms also involve Rab25 and has a clear impact on cancer behaviour. In fact, active integrins and CLIC3 are necessary for pancreatic cancer cell invasion.[49]

3. Adhesion Mediated Signalling in Colorectal Cancer

During tumor neo-angiogenesis, both soluble factors and fixed components of the ECM and membrane proteins determine signal exchange between the TME and the implicated cell types. The signalling network is coordinated by functional "hubs", which are constituted by integrin receptors, often associated with other proteins to form macromolecular signalling platforms at the adhesive sites.[50] For example, reciprocal interactions between integrin $\alpha v \beta 3$ and the VEGF-Receptor 2 (VEGF-R2) on endothelial cells are particularly important during tumor vascularization.[51] The mesenchymal aspects of angiogenesis, including endothelial cell invasion and vascular remodeling, rely on another integrin, the $\alpha 5 \beta 1$ integrin.[52] The latter turned out to drive angiogenesis in gliomas and to be one of the main regulators of tumor resistance to anti-angiogenesis treatment.[53]

What is new and relevant to the purposes of the present review is the fact that the integrin-centered angiogenesis hubs can include ion channels.[30] In general, ion channels and other transporters regulate the development of cancer hallmarks in different human tumors.[21,54] In particular, we found that the K^+ channels encoded by the human ether-à-go-go related gene (Kv11.1, or hERG1) are frequently overexpressed in human cancers.[54,55] hERG1 activity can regulate intracellular signalling by physically associating which integrin subunits and growth factor/chemokine receptors.[30,47,48,56] Therefore, it appears to be a central component of the molecular hubs that regulate neoangiogenesis in neoplasia.

3.1. *Integrin receptors, hERG1 channels and angiogenesis in colorectal cancer*

hERG1 channels are expressed in colorectal cancers (CRCs;[57]) and constitute a negative prognostic factor in nonmetastatic patients.[58] We recently provided evidence that hERG1 regulates tumor angiogenesis in CRCs by modulating the VEGF-A pathway.[48] We used a number of human CRC cell lines to determine whether VEGF-A secretion depends on functional hERG1 expression on the CRC cells' surface. VEGF-A secretion generally decreased when hERG1 activity was blocked by pharmacological inhibition or by down-regulation with anti-hERG1 small interfering RNAs (siRNA). The extent of inhibition is similar to the one obtained using an anti-*VEGF-A* siRNA. Moreover, blocking hERG1 was virtually ineffective in cell lines with low endogenous expression of the channel. In these cells, overexpressing hERG1 significantly increased VEGF-A secretion. These effects are specific for VEGF-A, as hERG1 inhibition does not affect secretion of other angiogenic factors, such as bFGF.[59] Considering that hERG1 activation tends to produce cell hyperpolarization,[60,61] it is unlikely that in CRC cells hERG1 directly stimulates VEGF-A secretion by activating Ca^{2+} influx through voltage-gated Ca^{2+} channels, which would require cell depolarization. Indeed, in those cells in which such mechanism is operant, like pancreatic beta cells[62] or chromaffin cells,[63] hERG1 block stimulates hormone secretion, contrary with the evidence we obtained in CRC cells. We thus hypothesized that the action of hERG1 on secretion in CRC cells is indirect. In particular, we found that the hERG1-dependent regulation of VEGF-A secretion mainly occurs by control of *VEGF-A* gene transcription. Because hERG1 channels can regulate intracellular signalling through a cross talk with integrin receptors containing the β1 subunit,[30] we tested the effects of either a β1-inhibiting (BV7) or a β1-activating (TS2/16) antibody on *VEGF-A* expression in CRC cells. *VEGF-A* expression is inhibited by BV7 or the hERG1 inhibitor E4031, whereas it is potentiated by TS2/16. Thus, β1 integrin appears to flank the effect of hERG1 on expression and secretion of VEGF-A.

3.2. *Cross-talk between the hERG1/β1 complex and the PI3K/Akt pathway*

Because hERG1 and β1 cooperate in regulating the angiogenic signalling in CRC cells, we also studied whether they physically associate in these models. Immuno-precipitation methods revealed that assembly of the hERG1/β1 complex depends on both integrin activation and proper functioning of hERG1 (as it is inhibited by blocking the channel). The Focal Adhesion Kinase (FAK) is also recruited into the complex. However, it displays only a slight phosphorylation on tyrosine 397 when associated with hERG1 and phosphorylation is independent of hERG1 activation. This result suggests that other functionally important signalling proteins may be associated with the hERG1/β1 complex or thereby activated. In brief, we observed that the membrane complex was always associated with the p85 subunit of PI3K, which turns out to be generally phosphorylated (i.e. activated) in CRC cells. Phosphorylation of p85 was strictly dependent on both β1-integrin and hERG1 activity. This mechanisms fed forward on Akt activity, which was found to be also dependent on hERG1 and integrin activation. On the other hand, other signalling molecules normally active in CRCs, such as ILK (integrin-linked kinase) and c-Src, were not affected by modulating hERG1 activity.

As recalled above, the *VEGF-A* transcription is regulated by HIF-1 and HIF-2. The effect is mediated by the Akt pathway, through mTOR,[64] FoxO[65] and p53.[66] We then further studied this pathway in CRC cells. We observed that VEGF-A secretion was decreased not only by inhibiting the PI3K/Akt pathway, but also by blocking hERG1 or silencing its expression. Moreover, the Akt1 isoform was much more implicated in the pathway, as compared to Akt2. These and other observations led us to conclude that the role of hERG1 inside the β1/hERG1/PI3K is the triggering of a signalling pathway which, through the involvement of Akt (mainly Akt1), activates HIF-1α and HIF-2α transcription. The ensuing increase in the HIF subunit concentration potentiates the transcription of HIF-dependent genes, including *VEGF-A*. Interestingly, we found no evidence of a mTOR-dependent control of HIF(s) protein synthesis, so that the mechanism resembles the angiotensin II-dependent regulation of HIF-1α and HIF-2α transcription in vascular smooth muscle cells.[67]

Moreover, the process is also independent of FoxO, but appears to involve p53. p53 is the product of the tumor suppressor gene *TP53*, which is commonly mutated in human cancers. p53 exerts several fundamental functions inside the cell, playing a pivotal role in the control of genomic stability as well as of cell survival. For these reasons, the amount and activity of p53 are normally kept under control, through the intervention of Mdm2 and of the nuclear protein p300. For the purposes of the mechanism described above, it is worth recalling that p53 also controls angiogenesis, since it modulates HIF protein(s) degradation.[68] The interplay between hERG1 and p53 merits further studies, as exemplified by the mathematical modelling reported below.

On the whole, we have identified a novel signaling pathway (outlined in Fig. 1), centered on the interaction between β1 integrins and hERG1 K$^+$ channels. The latter hence belong to the membrane functional hub which recruits and activates molecules which exert key signalling functions in cancer cell physiology.

3.3. *The* in vivo *evidence*

After determining *in vitro* the above signalling pathway, we looked into whether VEGF-A is regulated by hERG1 engagement *in vivo*, and it is relevant for tumor progression and the related angiogenesis. First, we studied the effects of injecting subcutaneously either HEK 293 cells overexpressing hERG1 or hERG1-silenced HCT116 cells into immunodepressed mice. In brief, it turned out that hERG1

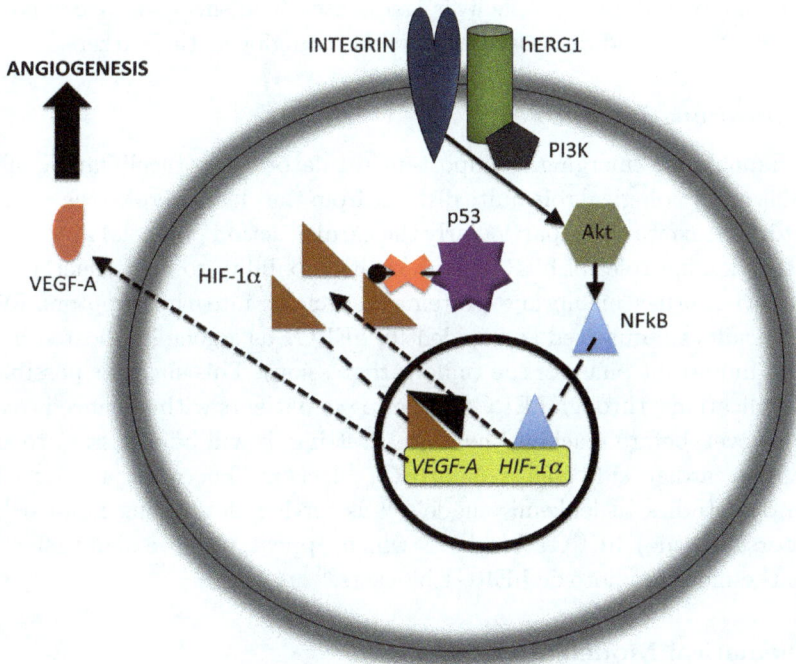

Fig. 1.   Scheme illustrating the hERG1 and β1 integrin interaction and the intracellular signalling pathway leading to control VEGF-A secretion. Note that when the levels of HIF-1α increase, the inhibitory effect of p53 is bypassed. For details, see text and Ref. 48.

expression strongly increased the volume of the cell masses produced in the animals as well as the VEGF-A expression and the induced vascular area. Next, we studied the effects of regulating hERG1 activity by treating the mice with hERG1 blockers. Treatment does not alter the behaviour, vitality or electrocardiographic properties.[47] However, blocking hERG1 is accompanied by a strong decrease of: (i) tumor volume; (ii) expression of VEGF-A, pAkt and Ki67; (iii) total vascular area. The tumor masses obtained from mice treated with hERG1 blockers display a decrease of pAkt and HIF-1α expression.

The effects of hERG1 blockers were also studied in an orthotopic CRC model.[48] HCT116 cells were injected into the coecum wall of immunodeficient mice and then treated or not with hERG1 blockers. Three months after the end of treatment, we found that control mice displayed macroscopic tumor masses of human origin, as expected, with numerous metastatic foci and complications such as ascites. In E4031-treated mice, the formation of tumor masses was strongly reduced and no evidence was present of distant metastases. The effect of blocking hERG1 was also tested in a model with liver metastases. HCT116 cells expressing luciferase were injected into the spleen of immunodeficient mice, from which they accessed the bloodstream. Spleens were removed the day after inoculum. When cells had reached the liver and grown there as single metastatic foci, mice were treated with E4031 for two weeks. Treatment delayed the metastatic process and strongly reduced the number of macro- and microscopic liver metastases. The effect was accompanied by larger necrotic areas and decreased expression of angiogenetic markers.

3.4. *Implications for oncology*

hERG1 channels are emerging as important regulators of intracellular signalling in tumor cells, a physiological role quite distinct from the classical roles these ion channels exert in cell excitability, particularly the cardiac action potential repolarization. The novel signalling roles of hERG1 turn on its capability to form macromolecular complexes with other membrane proteins, especially integrin receptors. What is more, the results summarized above identify hERG1 as a gene implicated in angiogenesis (an important phase of the tumor progression). This suggests possible therapeutic applications, through hERG1 targeting in patients with advanced colorectal cancer. However, before reaching the clinical setting, it will be necessary to exclude the potential cardiac side effects of hERG1 blockers. The approach we suggest, based on our studies in leukemic models,[47] is further developing non-cardiotoxic (i.e. nontorsadogenic) hERG1 blockers, which appear to be equally effective on tumor as the more cardiotoxic hERG1 blockers.[47]

4. Mathematical Modelling

Starting from this setting, it is important to study the interlaced dynamics of key microscopic actors implicated in the aforementioned processes, via dedicated mathematical models. In Ref. 69, a model was introduced to elucidate the coupled dynamics of apoptosis and hypoxia, with particular emphasis on the role played by angiogenesis. In the following we shall briefly review the model and discuss the main conclusion of our analysis.

In defining the model, we have preliminarily identified the molecules that provide an effective bridge between the two macroscopic processes under study, i.e. hypoxia/angiogenesis on one side and apoptosis on the other. These molecules are p53 and the p300 co-activator. A key role is however also played by oxygen, which provides an indirect coupling between the molecular pathways implicated in the above processes. The model also accounts for the evolution of the hypoxia inducible factor HIF-1, the caspases (CASP) and K^+ ions. Activation of CASP is a central process of apoptosis. When CASP are activated, the cell death program proceeds through rapid and sequential enzyme activations, which lead to disruption of the cytoskeleton and nuclear matrix. CASP dynamics is coupled to that of K^+ ions, which mimics the role of HERG channels. K^+ is involved in many cellular processes and presumably acts as a negative feedback on the apoptotic process. In fact, CASP activation is favored under low K^+ conditions.[70]

A scheme of the relevant interactions is given in Fig. 2, taken from Ref. 69. The stylized membrane delimits the inside of the cell, while two macro-compartments are depicted to cluster the processes and molecules implicated, respectively, in hypoxia/angiogenesis (dashed line) and apoptosis (solid line). The elements p53 and p300 take a part in both processes and provide the link among them. A similar consideration holds for the oxygen molecules, which are assumed to populate the surrounding environment, a sort of shared reservoir. As illustrated in Fig. 2, oxygen inhibits the formation of HIF-1, which is instead more abundant under hypoxia. HIF-1 binds p300 and stimulates the production of VEGF-A, which in turn triggers angiogenesis and thus a further increase in the oxygen income. Under hypoxic conditions p53 accumulates and competes with HIF-1 to bind p300. The activation of p53 causes a negative feedback to inhibit the formation of HIF-1, while seeding a cascade of reactions (not detailed into the model) which ultimately activate CASP. As mentioned above, CASP activation is instead opposed by the presence of intracellular K^+. Further, the decrease of oxygen inside the cell causes the loss of oxidative phosphorylation and reduced ATP production, which eventually leads to intracellular Na^+ accumulation ans increased extracellular K^+.[71] K^+ is therefore regulated, at least in part, by oxygen, a process that is accounted for by hypothesizing the existence of an effective and direct coupling between K^+ and O_2.

The scheme of interaction that we have put forward can be readily translated into the set of chemical equations reported in Fig. 3. Birth and death reactions (i)–(vii) control the constitutive expression of the molecules. These processes follow complex mechanisms, whose detailed representation is beyomd the scope of the present model. For this reason, they are imagined as generic creation and annihilation reactions controlled by dedicated reaction rates, which play the role of free parameters. Reaction (viii) involves oxygen and HIF-1 and results in the annihilation of the reactants. In other words, the presence of oxygen prevents the formation

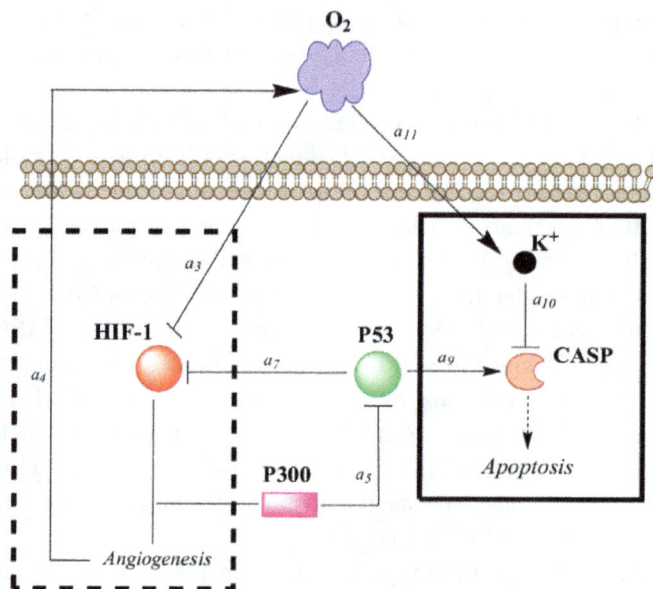

Fig. 2. The scheme of the model is depicted (see main text for an account of the processes implicated). For details concerning the reactions involved in the caspases activation via p53, see the maps Ko04115, KEGG database.[73] The angiogenesis network is contained as part of the large Ko05200 map on pathways in cancer. The interaction among p300 and HIF-1 is dicussed in references.[73,74] The mutual interference of p53 and p300 is object of analysis in Refs. 74 and 75. p53 is further assumed to inhibit HIF-1, as follows the study of e.g. Ref. 76.

(i) $\emptyset \xrightarrow{a_{hif}} \text{HIF-1}$

(ii) $\emptyset \xrightarrow{a_{o2}} O_2$

(iii) $\emptyset \xrightarrow{a_s} \text{p300}$

(iv) $\emptyset \xrightarrow{a_{p53}} \text{P53}$

(v) $\emptyset \xrightarrow{a_{12}} \text{CASP}$

(vi) $\text{CASP} \xrightarrow{a_{13}} \emptyset$

(vii) $\text{K} \xrightarrow{a_{14}} \emptyset$

(viii) $O_2 + \text{HIF-1} \xrightarrow{a_3} \emptyset$

(ix) $\text{HIF-1} + \text{p300} \xrightarrow{a_4} O_2$

(x) $\text{p300} + \text{p53} \xrightarrow{a_5} \emptyset$

(xi) $\text{p53} + \text{HIF-1} \xrightarrow{a_7} \text{p53}$

(xii) $\text{p53} \xrightarrow{a_9} \text{CASP}$

(xiii) $\text{CASP} + \text{K} \xrightarrow{a_{10}} \emptyset$

(xiv) $O_2 \xrightarrow{a_{11}} \text{K}$

Fig. 3.   The chemical reactions that defined the examined model.

of HIF-1. Indeed, the presence of oxygen destabilizes the alpha subunit of HIF-1 causing its elimination via the proteasome, and consequently inhibiting the formation of the HIF-1 complex. On the other hand, under hypoxic conditions the production of HIF-1 is stimulated. Then, HIF-1 binds p300 and, among the others, promotes the transcription of the *VEGF-A* gene. The latter, after translation and secretion, triggers angiogenesis: new blood vessels are created which enhance the oxygen income. This process can be ideally mimicked by chemical equation (ix).

Moreover, p300 is implicated in the degradation of p53, as accounted for by reaction (x). p53 and HIF-1 compete for the common target p300, an empirical fact that implies the existence of an indirect coupling between p53 and HIF-1 species. Under severe and prolonged hypoxia conditions an increase in the p53 concentration is in fact registered which eventually stimulates the degradation of the subunit HIF-1α, so inhibiting the formation of HIF-1[72] (via an interaction mediated by Mdm2). This process is effectively accounted for by equation (xi). As a further point, p53 induces the activation of the CASP[70] a process that is condensed in equation (xii). CASP need an environment with a low K^+ concentration to perform their enzymatic functions. One can therefore imagine that CASP get inhibited by K^+, a negative feedback that we implement through equation (xiii). Under hypoxia, the decrease of intracellular oxygen causes loss of oxidative phosphorylation and reduced ATP production, which in turn affects the functioning of the ATP-dependent sodium pump, thus altering the sodium and potassium distribution.[71] The presence of K^+ appears therefore to be partially regulated by the presence of oxygen, as dictated by equation (xiv).

Starting from the stochastic chemical model reported in Fig. 3, and applying the law of mass action, one can obtain a closed system of ordinary differential equations (ODEs), which governs the coupled evolution of the concentrations of the species involved. The mean field equations are given in Fig. 4. The reader can refer to Ref. 69 for further details on the derivation of the continuum deterministic limit, and to learn about the characterization of the equilibrium dynamics of the system defined in Fig. 3. In the remaining part of this section, we will instead focus on the main biological conclusion of our study. We will in particular highlight the presence of a dynamical switch between normoxia and cellular death conditions, a prediction of the reductionist model whose implications are discussed with respect to cancer progression.

To this end we recall that the relative concentration of CASP and K^+ is implicated in the process of apoptosis. The cell dies if CASP are over-expressed, while a large concentration of K^+ (hence low CASP amount) corresponds to healthy conditions. Working with the model defined by the system of equations in Fig. 3, it is possible to single out the regions of the relevant parameters space for which the CASP are predicted to prevail over K^+. To simplify the interpretation of the results, we assume that this latter condition identifies the region of cell death. The opposite condition, when K^+ takes over CASP, defines the domain where the cell is alive. All parameters are set to nominal values, except for a_{o2} and a_{p53}, respectively

$$\frac{d\mathcal{HIF}}{dt} = a_{hif} - a_3\mathcal{O}_2\mathcal{HIF} - a_4\mathcal{HIF}\mathcal{P}300 - a_7\mathcal{P}53\mathcal{HIF}$$

$$\frac{d\mathcal{O}_2}{dt} = a_{o_2} - a_3\mathcal{O}_2\mathcal{HIF} + a_4\mathcal{HIF}\mathcal{P}300 - a_{11}\mathcal{O}_2$$

$$\frac{d\mathcal{P}300}{dt} = -a_4\mathcal{HIF}\mathcal{P}300 - a_5\mathcal{P}300\mathcal{P}53 + a_8$$

$$\frac{d\mathcal{P}53}{dt} = -a_5\mathcal{P}53\mathcal{P}300 + a_{p_{53}} - a_9\mathcal{P}53$$

$$\frac{d\mathcal{CASP}}{dt} = a_9\mathcal{P}53 - a_{10}\mathcal{CASP}\mathcal{K}^+ + a_{12} - a_{13}\mathcal{CASP}$$

$$\frac{d\mathcal{K}^+}{dt} = -a_{10}\mathcal{CASP}\mathcal{K}^+ + a_{11}\mathcal{O}_2 - a_{14}\mathcal{K}^+$$

Fig. 4.   The deterministic equations for the concentration amounts. The variables are written in italic to recall that they are continuum concentrations.

associated with the oxygen production and $P53$ cycle, which can be freely tuned. In Fig. 3 of Ref. 69, a solid line is depicted which marks the marginal condition where the asymptotic (equilibrium) concentration of CASP is equal to the concentration of K^+. The line is calculated analytically for the limiting condition of equal concentration of HIF and p53. By integrating ODE system, one can appreciate the progressive transition from the region where CASP dominates (death region, according to our interpretation) towards the domain where K^+ takes over (life). The results of the numerical integration are displayed in Fig. 4 of Ref. 69, where the ratio $CASP/K^+$ is plotted for each selected pair (a_{o2}, a_{p53}), using a color-code representation. Both the calculations and the simulations can be extended beyond the limiting case study $HIF = P53$. This is accomplished by modulating the parameter a_{hif} beyond the specific value that yields the aforementioned condition (see Ref. 69 for details). In particular, by making a_{hif} larger, one detects a progressive downward shift of the transition line. The domain that we associate to life conditions gets smaller. The opposite holds when the reaction rate a_{hif} is conversely reduced. The transition line moves upward, widening the portion of the parameters plane that is associated to cell vitality. Cast in other terms, for any fixed pair (a_{o2}, a_{hif}), a critical value of a_{p53} exists that discriminates between death and life of a cell. This result, is summarized in Fig. 5 (taken from Ref. 69) and suggests an intriguing biological interpretation, on which we elaborate in the following paragraph.

Imagine that a cell experiences a reduction in the oxygen income, e.g. when a tumor is in the avascular phase. This situation can be described within the model setting by imposing a reduction of a_{o2}, while keeping the other parameters unchanged. As a consequence of this adjustment, the cell can fall on the other side of the transition line as depicted in Fig. 5, where the apoptosis condition are met. To oppose the death process, a cell can undergo a punctual mutation altering the rate of p53 production, i.e. reducing the parameter a_{p53}. Only the cells that are able to perform efficiently this task can escape from death, provoked by a

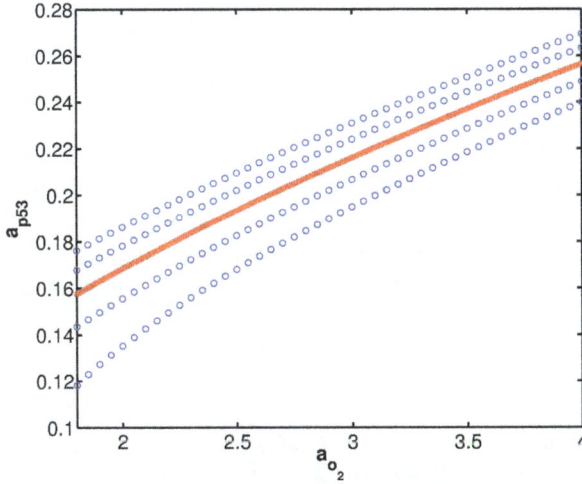

Fig. 5. The solid line in the plane (a_{o2}, a_{p53}) delineates the condition $CASP = K^+$, for $HIF = p53$. The symbols stand for the transition lines as obtained for values of a_{hif} for which the latter condition breaks. For larger values of a_{hif} the curves shift downward widening the region of cellular death. The opposite holds when a_{hif} is increased.

lower oxygen income. This could explain in turn why p53 is found to be mutated in more than 50% of human tumors, an ability that we interpret as instrumental in enhancing the survival chances of a cancer colony.

5. Conclusions

Increasing evidence points to a complex interplay between cancer cells and the microenvironment, which often determines tumor progression and the onset of chemoresistance. Such interplay is modulated by adhesion receptors of the integrin family, which often co-operates with ion channels expressed on the plasma membrane of cancer cells. We have detailed the interaction between a peculiar type of K^+ channel, hERG1, and β1-containing integrins. Moreover, we determined a novel signalling pathway, which is triggered by the integrin/hERG1 interaction and leads to the regulation of angiogenesis. This mechanism has a clear impact on cancer establishment and progression, as evidenced by the fact that cell proliferation is enhanced, whereas apoptosis is hampered when the integrin/channel complex is activated. This greatly impacts on the development of cancer metastases, and hence malignancy, in CRC cancer.

Cancer constitutes a complex system that involves different spatial and temporal scales and many molecular entities. Mathematical models can help to rationalize the dynamical interplay between distinct pathways and elucidate the role played by key actors. We have here reviewed a simple model of reductionist inspiration that explores the intimate connections between apoptosis and hypoxia, passing through angiogenesis. Monitoring the relative concentrations in CASP and K^+ ions, we have shown that specific alterations in the p53 production rate makes it possible for the cells to oppose the apopotic death, triggered by a low oxygen income. This could explain in turn why p53 is so relevant for establishment and progression of human cancers, as well the relationships of p53 regulation and K^+ channels activity, as indicated by experimental data.

On the whole, starting from these preliminary evidences, it appears to be mandatory to link experimental data and mathematical modelling to obtain proper understanding on the complex interplay between tumor cells and the TME. This could also have profound influence in designing new anticancer treatments.

References

1. P. M. Comoglio and L. Trusolino, *Nat. Med.* **11**, 1156 (2005).
2. L. Girieca and C. Ruegg, *Histochem. Cell. Biol.* **130**, 1091 (2008).
3. J. A. Joyce and J. W. Pollard, *Nat. Rev. Cancer* **9**, 239 (2009).
4. F. Nwajei and M. Konopleva, *Adv. Hematol.* **2013**, 953982 (2013).
5. D. Mahadevan and D. D. Von Hoff, *Mol. Cancer Ther.* **6**, 1186 (2007).
6. D. Hanahan and R. A. Weimberg, *Cell* **144**, 646 (2011).
7. E. Ruoslahti, S. N. Bhatia and M. J. Sailor, *J. Cell Biol.* **188**, 759 (2010).
8. D. Hanahan and J. Folkman, *Cell* **86**, 353 (1996).
9. I. J. Fidler and L. M. Ellis, *Cell* **79**, 185 (1994).
10. N. Ferrara, H. P. Gerber and J. LeCouter, *Nat. Med.* **9**, 669 (2003).
11. J. A. Forsythe, B.-H. Jiang, N. V. Iyer, F. Agani, S. W. Leung, R. D. Koos and G. L. Semenza, *Mol. Cell. Biol.* **16**, 4604 (1996).
12. F. Dayan, N. M. Mazure, C. Brahimi-Horn and J. A. Pouyssegur, *Cancer Microenv.* **1**, 53 (2008).
13. D. Liao and R. S. Johnson, *Cancer Met. Rev.* **26**, 281 (2007).
14. E. B. Rankin and A. J. Giaccia, *Cell Death Differ.* **15**, 678 (2008).
15. L. E. Huang, J. Gu, M. Schau and H. F. Bunn, *P. N. A. S. USA* **95**, 7987 (1998).
16. G. L. Semenza, *Wiley Interdiscip. Rev: Syst. Biol. Med.* **2**, 336 (2009).
17. T. Imamura, H. Kikuchi, M. T. Herraiz, D. Y. Park, Y. Mizukami, M. Mino-Kenduson, M. P. Lynch, B. R. Rueda, Y. Benita, R. J. Xavier and D. C. Chung, *Int. J. Cancer* **124**, 763 (2009).
18. D. Feldser, F. Agani, N. V. Iyer, B. Pak, G. Ferreira and G. L. Semenza, *Cancer Res.* **95**, 3915 (1999).
19. A. J. Giaccia, G. S. Bronwyn and S. J. Randall, *Nat. Rev. Drug Discov.* **2**, 803 (2003).
20. A. Arcangeli, *Am. J. Physiol.* **30**, C762 (2011).
21. A. Arcangeli, O. Crociani, E. Lastraioli, A. Masi, S. Pillozzi and A. Becchetti, *Curr. Med. Chem.* **16**, 66 (2009).
22. A. Arcangeli and A. Becchetti, *Pharmaceuticals* **3**, 1202 (2010).
23. M. D'Amico, L. Gasparoli and A. Arcangeli, *Recent Pat. Anticancer Drug Discov.* **8**, 53 (2013).
24. J. M. de Pereda, G. Wiche and R. C. Liddington, *EMBO J.* **18**, 4087 (1999).
25. R. O. Hynes, *Cell* **110**, 673 (2002).
26. S. Tadokoro, S. J. Shattil, K. Eto, V. Tai, R. C. Liddington, J. M. de Pereda, M. H. Ginsberg and D. A. Calderwood, *Science* **302**, 103 (2003).
27. M. H. Ginsberg, A. Partridge and S. J. Shattil, *Curr. Opin. Cell. Biol.* **17**, 509 (2005).
28. C. K. Miranti and J. S. Brugge, *Nat. Cell. Biol.* **4**, E83 (2002).

29. S. Cabodi, P. Di Stefano, M. del Pilar Camacho Leal, A. Tinnirello, B. Bisaro, V. Morello, L. Damiano, S. Aramu, D. Repetto, G. Tornillo and P. Defilippi, *Adv. Exp. Med. Biol.* **674**, 43 (2010).
30. A. Arcangeli and A. Becchetti, *Trends Cell. Biol.* **16**, 631 (2006).
31. A. Arcangeli, A. Becchetti, A. Mannini, G. Mugnai, P. Defilippi, G. Tarone, M. R. Del Bene, E. Barletta, E. Wanke and M. Olivotto, *J. Cell. Biol.* **122**, 1131 (1993).
32. A. Becchetti, A. Arcangeli, M. R. Del Bene, M. Olivotto and E. Wanke, *Proc. R. Soc. Lond. B. Biol. Sci.* **248**, 235 (1992).
33. P. Doherty, S. V. Ashton, S. E. Moore and F. S. Walsh, *Cell* **67**, 21 (1991).
34. G. Hofmann, P. A. Bernabei, O. Crociani, A. Cherubini, L. Guasti, S. Pillozzi, E. Lastraioli, S. Polvani, B. Bartolozzi, V. Solazzo, L. Gragnani, P. De Filippi, B. Rosati, E. Wanke, M. Olivotto and A. Arcangeli, *J. Biol. Chem.* **276**, 4923 (2001).
35. R. Belusa, O. Aizman, R. M. Andersson and A. Aperia, *Am. J. Physiol. Cell. Physiol.* **282**, C302 (2002).
36. R. Menegazzi, S. Busetto, R. Cramer, P. Dri and P. Patriarca, *J. Immunol.* **165**, 4606 (2000).
37. T. Tominaga and D. L. Barber, *Mol. Biol. Cell.* **9**, 2287 (1998).
38. R. A. Cardone, V. Casavola and S. J. Reshkin, *Nat. Rev. Cancer* **5**, 786 (2005).
39. M. D. Cahalan and K. G. Chandy, *Immunol. Rev.* **231**, 59 (2009).
40. M. J. Davis, X. Wu, T. R. Nurkiewicz, J. Kawasaki, P. Gui, M. A. Hill and E. Wilson, *Cell. Biochem. Biophys.* **36**, 41 (2002).
41. A. Katsumi, A. W. Orr, E. Tzima and M. A. Schwartz, *J. Biol. Chem.* **279**, 12001 (2004).
42. M. Levite, L. Cahalon, A. Peretz, R. Hershkoviz, A. Sobko, A. Ariel, R. Desai, B. Attali and O. Lider, *J. Exp. Med.* **191**, 1167 (2000).
43. V. V. Artym and H. R. Petty, *J. Gen. Physiol.* **120**, 29 (2002).
44. A. Cherubini, S. Pillozzi, G. Hofmann, O. Crociani, L. Guasti, E. Lastraioli, S. Polvani, A. Masi, A. Becchetti, E. Wanke, M. Olivotto and A. Arcangeli, *Ann. N. Y. Acad. Sci.* **973**, 559 (2002).
45. A. Cherubini, G. Hofmann, S. Pillozzi, L. Guasti, O. Crociani, E. Cilia, M. Balzi, S. Degani, P. Di Stefano, P. Defilippi, R. Wymore and A. Arcangeli, *Mol. Biol. Cell.* **16**, 2972 (2005).
46. S. Pillozzi, M. F. Brizzi, P. A. Bernabei, B. Bartolozzi, R. Caporale, V. Basile, V. Boddi, L. Pegoraro, A. Becchetti and A. Arcangeli, *Blood* **110**, 1238 (2007).
47. S. Pillozzi, M. Masselli, E. De Lorenzo, B. Accordi, E. Cilia, O. Crociani, A. Amedei, M. Veltroni, M. D'Amico, G. Basso, A. Becchetti, D. Campana and A. Arcangeli, *Blood* **117**, 902 (2011).
48. O. Crociani, F. Zanieri, S. Pillozzi, E. Lastraioli, M. Stefanini, A. Fiore, A. Fortunato, M. D'Amico, M. Masselli, E. De Lorenzo, L. Gasparoli, M. Chiu, O. Bussolati, A. Becchetti and A. Arcangeli, *Sci. Rep.* **3**, 3308 (2013).
49. M. A. Dozynkiewicz, N. B. Jamieson, I. Macpherson, J. Grindlay, P. V. van den Berghe, A. von Thun, J. P. Morton, C. Gourley, P. Timpson and C. Nixon, *Dev. Cell.* **22**, 131 (2012).
50. E. J. Brown, *Curr. Opin. Cell Biol.* **14**, 603 (2002).
51. P. R. Somanath, N. L. Malinin and T. V. Byzova, *Angiogenesis* **12**, 177 (2009).
52. S. Kim, S. Bell, S. Mousa and J. Varner, *Am. J. Pathol.* **156**, 1345 (2000).
53. A. Jahangiri, M. A. Aghi and S. Carbonell, *Cancer Res.* **74**, 3 (2014).
54. A. Becchetti and A. Arcangeli, *J. Gen. Physiol.* **132**, 313 (2008).
55. A. Arcangeli, Expression and Role of hERG Channels in Cancer Cells, in *The hERG Cardiac Potassium Channel: Structure, Function and Long QT Syndrome*, Novartis Foundation Symposium, eds. Chadwick D. J. and Goode J., p. 225 (2005).

56. S. Pillozzi and A. Arcangeli, *Adv. Exp. Med. Biol.* **674**, 55 (2010).
57. E. Lastraioli, L. Guasti, O. Crociani, S. Polvani, G. Hofmann, H. Witchel, L. Bencini, M. Calistri, L. Messerini, M. Scatizzi, R. Moretti, E. Wanke, M. Olivotto, G. Mugnai and A. Arcangeli, *Cancer Res.* **15**, 606 (2004).
58. E. Lastraioli, L. Bencini, E. Bianchini, M. R. Romoli, O. Crociani, E. Giommoni, L. Messerini, S. Gasperoni, R. Moretti, F. Di Costanzo, L. Boni and A. Arcangeli, *Transl. Oncol.* **5**, 105 (2012).
59. A. Masi, A. Becchetti, R. Restano-Cassulini, S. Polvani, G. Hofmann, A. M. Buccoliero, M. Paglierani, B. Pollo, G. L. Taddei, P. Gallina, N. Di Lorenzo, S. Franceschetti, E. Wanke and A. Arcangeli, *Brit. J. Canc.* **93**, 781 (2005).
60. M. C. Sanguinetti, C. Jiang, M. E. Curran and M. T. Keating, *Cell* **81**, 299 (1995).
61. R. Schönherr, B. Rosati, S. Hehl, V. G. Rao, A. Arcangeli, M. Olivotto, S. H. Heinemann and E. Wanke, *Eur. J. Neurosci.* **11**, 753 (1999).
62. B. Rosati, P. Marchetti, O. Crociani, M. Lecchi, R. Lupi, A. Arcangeli, M. Olivotto and E. Wanke, *Faseb J.* **14**, 2601 (2000).
63. F. Gullo, E. Ales, B. Rosati, M. Lecchi, A. Masi, L. Guasti, M. F. Cano-Abad, A. Arcangeli, M. G. Lopez and E. Wanke, *Faseb J.* **17**, 330 (2003).
64. S. C. Land and A. R. Tee, *J. Biol. Chem.* **282**, 20534 (2007).
65. Y. Zhang, B. Gan, D. Liu and J. H. Paik, *Cancer Biol. Ther.* **12**, 253 (2011).
66. A. Sermeus and C. Michiels, *Cell Death Dis.* **2**, e164 (2011).
67. E. L. Pagé, D. A. Chan, A. J. Giaccia, M. Levine and D. E. Richard, *J. Biol. Chem.* **277**, 48403 (2007).
68. R. Ravi, B. Mookerjee, Z. M. Bhujwalla, C. H. Sutter, D. Artemov, Q. Zeng, L. E. Dillehay, A. Madan, G. L. Semenza and A. Bedi, *Genes Dev.* **14**, 34 (2000).
69. P. Laise, D. Fanelli and A. Arcangeli, *Commun. Nonlinear Sci. Numer. Simul.* **17**, 1795 (2012).
70. F. M. Hughes and J. A. Cidlowski, *Adv. Enzyme Regul.* **39**, 157 (1999).
71. R. S. Contran, V. Kumar and T. Collins, *Robbins, Pathologic Basis of Disease*, 6th edn. (W.B. Saunders Company, Philadelphia, 1999).
72. T. Schmid, J. Zhou, R. Kohl and B. Brune, *Biochem. J.* **380**, 289 (2004).
73. M. Kanehisa and S. Goto, *Nucleic Acids Res.* **28**, 27 (2000).
74. H. M. Chan and N. B. La Thangue, *J. Cell Sci.* **114**, 2363 (2001).
75. S. R. Grossman, M. Perez, A. L. Kung, M. Joseph, C. Mansur, Z. X. Xiao, S. Kumar, P. M. Howley and D. M. Livingston, *Mol. Cell.* **2**, 405 (1998).
76. T. Schmid, J. Zhou, R. Kohl and B. Brune, *Biochem. J.* **380**, 289 (2004).

Chapter 8

Common and Diverging Integrin Signals Downstream of Adhesion and Mechanical Stimuli and Their Interplay with Reactive Oxygen Species

Kathrin Stephanie Zeller

Department of Immunotechnology
Lund University, Medicon Village, Building 406
223 81 Lund, Sweden
Kathrin.Zeller@immun.lth.se

Staffan Johansson

Department of Medical Biochemistry and Microbiology
Uppsala University, BMC, Box 582
751 23 Uppsala, Sweden
Staffan.Johansson@imbim.uu.se

The integrin family of adhesion receptors regulates basic functions of cells, and the signals they induce are altered in tumor cells. In this review we discuss how different integrin-dependent signals are generated during cell adhesion and by physical forces acting on cells. We also describe how reactive oxygen species are integral parts of integrin signaling and highlight a few important questions in the field. Answers to those may improve our understanding of integrins and their role in the development of cancer.

1. Introduction

The integrin family of adhesion receptors has crucial roles for numerous processes such as organ development, angiogenesis, blood coagulation, and immune responses. At the cellular level integrins contribute to these processes through their structural and signaling functions. They are the central components of hemidesmosomes, focal contacts and other cell contacts, and they regulate extracellular matrix formation. The signaling reactions induced by integrins are essential for several basic cellular functions, including survival, cell cycle regulation, and migration.[1,2] Integrin signals are generated by different types of cellular stimuli, i.e. ligand binding and various types of physical forces.[3,4] Importantly, the "integrin signals" triggered by these stimuli are not the same.[5]

Tumor development is known to depend on integrin-mediated adhesion and physical stimuli in several ways, which concern both the cancer cells themselves and the non-transformed host cells. Cell proliferation is regulated by cooperative signals from growth factors and integrins, e.g. in the activation of the ERK pathway and the passage through the G1/S checkpoint of the cell cycle.[6] Integrin adhesion also controls the cytokinesis process at the end of the cell cycle by poorly understood signals.[7] Integrins have been reported to cooperate with growth factor receptors by different mechanisms, including interactions in common receptor complexes, generation of intermediates necessary for the growth factor pathway (converging pathways), and amplification of growth factor signals.[8] Integrins may therefore contribute to the deregulated proliferation of cancer cells caused by defects in reactions downstream of integrins or several other receptors. Invasive growth and metastasis require cell migration, which depends on membrane protrusions at the cell front driven by actin polymerization and detachment at the rear driven by actin-myosin contraction. Both reactions are potently induced by integrins.[2,9] The invasive phenotype in carcinomas is linked to TGFβ-induced epithelial-mesenchymal transition, and the activation of the latent TGFβ complex occurs mainly via binding to integrin $\alpha v\beta 6$ in carcinomas.[10] The high TGFβ activity will in addition have other effects promoting tumor growth, including the formation of a collagen-rich stiff extracellular matrix (ECM) by resident fibroblast-like cells. Host cells contribute to tumor progression also by forming new blood vessels, whose angiogenesis requires integrins both for the migration of endothelial cells and pericytes, and for the remodeling of the ECM.

In order to better understand how integrin-dependent reactions are used by cancer cells to promote invasive growth, the ability to metastasize and to resist apoptosis-inducing conditions, the effects of defined integrin stimuli on cell signals need to be identified. These efforts include characterization of (i) responses induced by the different integrin stimuli, and (ii) integrin type-specific signals.

2. Integrin Signaling Mechanisms

Integrin signaling has mainly been studied in re-adhesion assays where suspended cells are seeded on immobilized integrin ligands. During the attachment and cell spreading phases in such assays many signaling reactions occur as extensively documented, e.g. the activation of focal adhesion kinase (FAK), phosphatidylinositol 3-kinase (PI3K), Rho GTPases, and the pathway components downstream of these proteins. However, the initial triggering events for the reactions are still poorly understood. Ligand binding to integrins results in large conformational changes in the extracellular domain,[11] but it is not known whether they are propagated across the plasma membrane. While this possibility should not be excluded, it has been shown that mere integrin clustering by polyvalent antibodies can induce recruitment and activation of cytoplasmic enzymes.[3,12] The understanding of the signaling mechanism(s) is also complicated by the later emerged role of integrins as mechano-receptors, whereby signals are generated through conformational changes in force-sensitive proteins associated with integrins.[4]

2.1. *Ligand-induced integrin signaling*

Cell attachment and spreading are driven by dynamic and incompletely understood reactions involving ligand-integrin interactions, force from actin polymerization pushing against the plasma membrane, and myosin-dependent pulling force on adhesion sites. Thus it is often unclear by which mechanism the signaling proteins were activated in published studies of re-adhering cells. Using inhibitors of myosin II (Blebbistatin) and RhoA kinase (Y27632) we found that several phosphorylation reactions during the initial stages of fibroblast attachment (< 30 min) to fibronectin occur independently of intracellular contractile forces.[5] Thus, mere ligand binding to integrin $\alpha 5\beta 1$ was sufficient to trigger phosphorylation of FAK-Y397, ERK-T202/Y204, AKT-S473, p130CAS-Y410, myosin phosphatase targeting subunit 1 (MYPT)-T853, myosin light chain (MLC)-S19 and cofilin-S3. This was the case also for the reactions driving integrin-dependent actin polymerization. The polymerization was monitored as lamellipodia protrusion from the initial contact points of attaching cells by live cell total internal reflection fluorescence (TIRF) microscopy, and this assay could detect the response to integrin ligand binding as early as after a few seconds.[5,9] Possibly, the contribution of myosin-dependent contractile force becomes important at later time points or affects other signaling reactions.

2.2. *Tension-induced integrin signaling*

Integrins act as mechano-sensors by linking extracellular ligands to actin filaments via adaptor proteins in adhesion sites.[4] Some integrin adhesion site-associated proteins can change conformation upon mechanical stimulation and thereby expose cryptic binding or phosphorylation sites. So far only few intracellular proteins have been clearly shown to be force-regulated, such as talin,[13] p130CAS,[14] and filamin,[15] but this may change if the methodological difficulties to study such conformational changes are overcome. Mechanical force has also been reported to regulate the structure and function of integrins themselves,[16] an interesting finding that is important to be investigated further. The forces acting on cells in our body include gravity, stretching by muscle work (breathing, pressure pulses from heart beats, pulling on tendons, etc.), shear stress from liquid flow, and contractile force generated by myosin II inside the cell. These physical stimuli are necessary for the development and maintenance of our body.[17,18] The signaling outcome of intracellular contraction is dependent on the stiffness of the surrounding ECM, and ECM stiffness may be a dominant factor for stem cell differentiation.[19] Tumor development has been reported to be affected by the tissue stiffness mainly in two ways: (i) soft ECMs foster selection of tumor initiating cells ("tumor stem cells") by induction of pluripotency genes, (ii) a stiff ECM (typical for solid tumors) promotes tumor growth and migration.[20]

A variety of approaches have been used to study the role of mechanical force on cell signaling reactions, and the reported results vary considerably. The cell type studied and how the force is applied (static, cyclic, frequency, amplitude, duration) are obvious factors that will influence the outcome of the experiments. For example, cyclic stretching for hours has been reported to affect the cytoskeletal organization,[21] oxidative stress levels,[22] and mRNA synthesis,[23] responses that will have many secondary effects in the cell. Some responses are seen within 5 to 10 min of mechanical stimulation,[5,23] and they are likely to be relatively direct results of the conformational changes in the force-sensitive proteins. Other factors that can affect the results are cell density and ion channel expression, information usually not provided in the published reports. Besides integrins, cadherins and several ion channels are believed to be the main mechano-receptors on cells.[24,25] The contributions of mechano-signals from cell-cell contacts can be analyzed and controlled by performing the experiments at low and high cell densities. However, the involvement of mechano-stimulated ion channels is presently more difficult to study. This is due to a lack of specific inhibitors and to the large number of different channels that makes knock-out and knock-down approaches complicated. It should also be noted that some ion channels have been reported to actually interact with and to be regulated by integrins.[26]

We have recently shown that short-term cyclic stretching (10 to 30 min) of sparsely seeded fibroblasts triggers activation of only a small number of integrin-associated signaling proteins compared to the signals generated during cell attachment (integrin ligation). The phosphorylation of ERK1/2 appears to be a particular stretch-responsive signal.[5] FAK has been suggested to become activated by unfolding in response to force and to be required for tension-induced activation of ERK.[27,28] However, consistent with the absence of force-induced FAK activation in our studies, phosphorylation of ERK1/2 was induced to the same degree by cyclic stretching (1 Hz) of FAK-null and FAK-expressing fibroblasts.[29] The question how this MAPK is activated by tensional force is intriguing and presently not understood as illustrated by the following examples: different reports have suggested that ERK activation by integrin-mediated tension correlates with or is dependent on activation of Ras,[30] inactivation of Ras,[31] influx of Ca^{2+},[24] or release of reactive oxygen species (ROS) from mitochondria.[22]

3. Integrin Type-Specific Signals

In spite of the vast literature on integrin-mediated signaling, the differences in responses between different integrins are still poorly characterized. Most cells express several integrin α and β subunits, and some of the resulting integrin heterodimers have overlapping ligand specificities. Therefore, in order to receive significant data regarding signaling specificity of the different integrins, cell systems and ligands need to be well characterized and controlled. The signals generated by $\beta 1$, $\beta 2$, or $\beta 3$ integrin subunits have been extensively studied while less is known about $\beta 4$ to $\beta 8$. The information regarding contributions from the α subunits to integrin signals is scarce. $\beta 1$ and $\beta 3$ integrins are expressed by most cultured adherent cell lines and can trigger similar signaling reactions. Yet, important functional outcomes are known to differ between the two fibronectin receptors $\alpha5\beta1$ and $\alpha v\beta3$. Integrin $\alpha v\beta3$ promotes the formation of large adhesion sites.[32] In contrast, $\alpha5\beta1$ induces the formation of smaller and more dynamic adhesion sites at the cell periphery.[32,33] and stronger traction force.[32,34] It is also much more efficient than $\alpha v\beta3$ in inducing fibronectin polymerization on the cell surface.[35] The underlying mechanisms for these actomyosin-related differences are not clear. However, myosin-II-dependent contraction requires activating phosphorylation on myosin light chain (MLC)-S19, and $\alpha5\beta1$ was recently shown to induce this phosphorylation more efficiently than $\alpha v\beta3$.[5,32]

The regulation of MLC phosphorylation involves at least two signaling pathways, i.e. RhoA/ROCK-dependent inactivation of MLC phosphatase and phospholipase C (PLC)/Ca^{2+}/calmodulin-dependent activation of MLC kinase.[36] Interestingly, the RhoA activity can be suppressed by $\beta3$-associated Src catalyzing an activating phosphorylation of RhoA GAP,[37] while the $\beta1$ cytoplasmic domain does not bind Src.[38] This is consistent with lower RhoA activity after adhesion mediated by $\alpha v\beta3$ compared to $\alpha5\beta1$.[39] However, $\alpha v\beta3$ has also been reported to induce a much higher RhoA activity than $\alpha5\beta1$ in adhesion assays[32]; since the traction force in this study was lower from $\alpha v\beta3$ than from $\alpha5\beta1$ and therefore did not correlate with the RhoA activities, it was concluded that only $\alpha5\beta1$ was able to promote coupling of RhoA to ROCK activation in these cells. However, we found that $\alpha5\beta1$ and $\alpha v\beta3$ induced the inactivating phosphorylation of MLC phosphatase (a direct measure of ROCK activity) equally efficiently in two different cell lines.[5] The varying results regarding the correlation of RhoA with MLC activity suggest that the activation

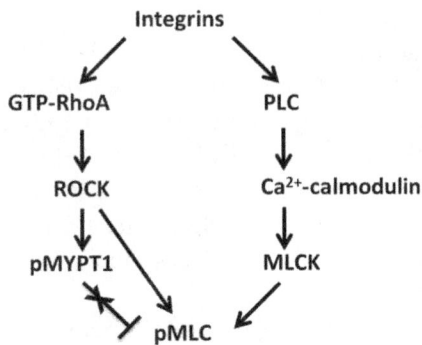

Fig. 1.   Phosphorylation of MLC by MLC kinase (MLCK) or ROCK activates myosin II to pull on actin filaments. ROCK also phosphorylates the MLC phosphatase subunit MYPT1 and thereby inhibits the inactivation of myosin II. ROCK can be activated by integrins via GTP-loaded RhoA.[30] MLCK activation requires Ca^{2+}-loaded calmodulin and is promoted by (locally) elevated cytoplasmic Ca^{2+} levels, which can be induced by integrin-activated phospholipase C (PLC).[40] The activating steps upstream of RhoA GEFs (including p190RhoGEF, p115, GEF-H1, and LARG)[30] and PLC triggered by integrins are incompletely understood. Adhesion via integrin $\alpha5\beta1$ is reported to induce phosphorylation of MLC more efficiently than adhesion via $\alpha v\beta3$.[5,32]

via $PLC/Ca^{2+}/calmodulin/MLC$ kinase has a dominating role in the regulation of myosin II and cell contraction. The regulation of MLC kinase by integrins (Fig. 1) therefore deserves more detailed studies.

4. Integrins and Reactive Oxygen Species

It has become increasingly clear that integrin functions are intimately associated with ROS action. Integrin stimulation generates significant amounts of ROS, and ROS affect integrin signals as well as a variety of other functions in cells. While it is important to realize that ROS is a summary name for very different molecules, each with characteristic properties and reactivity such as hydrogen peroxide, hydroxyl and superoxide radicals, cell signaling reactions are thought to be affected mainly by hydrogen peroxide due to its relatively long half-life. ROS are produced by several cellular oxidases, e.g. complexes in the mitochondrial electron transport chain, NADPH oxidases (NOXes) and 5-lipoxygenase (5-LOX) in the contexts of cell metabolism, pathogen defense and cell signaling.[41] ROS production is tightly controlled and defense systems limit ROS to act mainly in a localized fashion. Thereby potentially harmful effects of these promiscuously reactive molecules are prevented or reduced. Elevated ROS production from mitochondria and NOXes are commonly found in tumor cells and may be linked to increased migration and apoptosis resistance.[42,43]

Hydrogen peroxide preferentially targets redox-sensitive residues in proteins, cysteines being the prototypic example.[44] Quite early on it has been realized that cysteines in proteins often play roles for their enzymatic function, and more recently it has been acknowledged that the reversible modification of these residues by oxidation or conjugation is part of signal transduction mechanisms.[44,45] An important example is the reaction of hydrogen peroxide with phosphatases, e.g. the inhibition of PTEN.[45] which conserves the 3'-phosphorylated phosphatidylinositols and leads to sustained activities downstream of PI3Ks. Other targets include transcription factors,[41,46] kinases (e.g. Src[47]), small GTPases,[48-52] matrix metalloproteinases,[41,53] actin[54,55] and actin-associated molecules.[56-58] ROS acting on these targets will give rise to numerous feedback and feedforward reactions and is likely responsible for providing signal amplification and mediating cross-talk between different signal cascades.[59]

In order to control ROS effects, cells contain multiple protection systems such as ROS-converting enzymes (superoxide dismutases, catalase, glutathione peroxidase, peroxiredoxins) and other scavengers (e.g. GSH-GSSG, ascorbate), but there may be differences in how effectively these mechanisms work in different cell types and even in individual cells among one population. Additionally, different cellular compartments show marked differences in their redox potential.[44] These variations, together with the complicated chemistry and fast reactivity of ROS, underlie the complexity of this research field and the experimental difficulties it is facing. More detailed information can be found in several comprehensive reviews on ROS chemistry and biology.[41,60–65]

ROS from several sources have been linked to integrin engagement and signaling, both during attachment and mechanical stimulation of cells. Mitochondria-derived ROS[52,66] as well as ROS from NOXes and 5-lipoxygenase (5-LOX) produced in response to integrin ligand binding[67,68] were reported to affect cell attachment, spreading, and associated cytoskeletal changes. There are also indications that FAK, important for survival and migration, is regulated by ROS in response to integrin-mediated adhesion e.g. through the reversible oxidation (i.e. inhibition) of the phosphatases LMW-PTP[67,69] and SHP-2.[66] The inhibition of these phosphatases may allow a sustained phosphorylation and activation of FAK and thus the propagation of integrin signals. Less is known about ROS production downstream mechanical cell stretching, but NOXes[70,71] and mitochondria[22] as well as cross-activation between these sources[72] have been implicated. For example, Ali *et al.*[22] reported an increased FAK phosphorylation at Y397 in endothelial cells in response to cyclic strain, which could be abolished by antioxidants and mitochondrial inhibitors. However, several of the results on the role of ROS during different integrin stimuli were obtained with reagents that have poor specificity (e.g. diphenylene iodonium (DPI), apocynin, and N-acetylcysteine) or may cause artifacts (e.g. dyes such as 2,7-dihydrodichlorofluorescein diacetate (H_2DCFDA)).[73–75] Considering the variations in experimental settings in different studies, such as stimuli parameters, measurement methods and endogenous differences in cell lines and types, both with regard to ROS production and antioxidant capacities, general conclusions are difficult to draw at present.

In our own studies, we have obtained evidence that mitochondrial ROS affect AKT and ERK1/2 signaling pathways in two different fibroblast cell lines in re-adhesion assays (30 min after seeding).[5] Rotenone, an inhibitor of complex I in the respiratory chain with no known other targets, reduced AKT phosphorylation levels, what would be consistent with a higher activity of PTEN (and other phosphatases) when ROS released from mitochondria is inhibited. In this context it is interesting to note that AKT2 has been reported to co-localize with mitochondria[76] and that $\beta1$ integrins in MCF-7 cells preferentially activate the AKT2 isoform during attachment and spreading.[77] Also Taddei *et al.*[66] reported data supporting the importance of mitochondrial ROS in an early phase of cell attachment. The mechanism for how integrins transfer signals to mitochondria remains unclear, although both physical coupling via actin filaments and diffusible factors have been suggested.[22,52,78]

We also found that extracellular addition of catalase (a highly specific enzyme catalyzing the reaction from hydrogen peroxide to water and oxygen) enhances the stretch-induced ERK1/2 phosphorylation. A stable vitamin C derivate that is converted to active vitamin C (superoxide scavenger) by cellular enzymes inside and outside the cells has strikingly similar effects. These results, summarized in Table 1, point to a role of NOXes in stretch-induced signaling (see Ref. 5 and Fig. 2).

Table 1. Diverging effects of ROS derived from different cellular sources on signals during cell attachment and stretching.

	AKT-pS473		ERK-T202/Y204		p130CAS-Y410	
	BJ hTERT	GD25β1	BJ hTERT	GD25β1	BJ hTERT	GD25β1
Attachment + rotenone	−	−	−	+	=	=
Stretching + rotenone	=	=	=	=	=	=
Attachment + catalase	=	=	=	=	=	=
Stretching + catalase	=	=	+	+	=	=

= no change
− reduced phosphorylation
+ increased phosphorylation

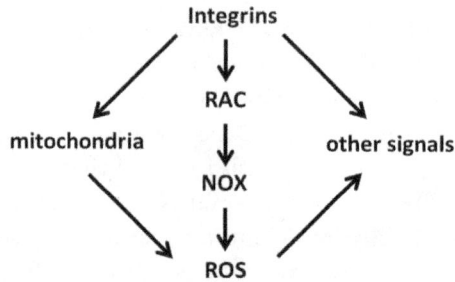

Fig. 2. Integrin stimulation generates ROS from different sources. Elevated release of hydrogen peroxide from mitochondria during cell attachment is induced by unknown mechanisms. Our data suggests that mechano-stimulation generates superoxide by NOXes located at the plasma membrane[5] and, possibly, in the case of NOX4, also at intracellular sites.[79–81] Superoxide can rapidly dismutate to hydrogen peroxide, which can pass through membranes. Integrin-mediated activation of NOX1 and NOX2 involves the activation of RAC, but the mechanisms for RAC activation as well as other steps in NOX activation are incompletely characterized. The generated ROS will significantly modulate the signaling reactions downstream of integrins as well as signals induced by other receptors.

In order to obtain more informative data, better methods allowing for both temporal and spatial resolution are needed. Several new methods to monitor certain ROS types or redox states have been developed employing for example boronate-based H_2O_2-selective probes (e.g. PeroxyGreen[82,83]), the genetically encoded H_2O_2 biosensor HyPer[84] and redox-sensitive GFP (for example roGFP[85,86]); all have been used for live cell measurements. However, it is important to choose the probes carefully depending on the research question and to ensure suitable conditions with appropriate controls in order to be able to draw valid conclusions.[87] For example, a measurement of the redox state does not provide relevant information about ROS concentrations. It will be interesting to follow if these promising probes work as hoped for, and to see if they can verify previous observations and provide opportunities to better understand the interplay between integrins and ROS.

5. Outlook

In spite of the vast amount of research that has been performed in the integrin field during the past decades, there are several key questions remaining to clarify.

(i) Integrins: What are the signals deriving from distinct members of the integrin family during ligand binding? And what are the roles of the integrin α units in signaling? We still have not yet clearly revealed if integrins themselves are mechano-sensitive, i.e. if their conformation or clustering is affected by force, and if it makes a difference if the force comes from inside or outside of the cells.

(ii) Force-induced signals: It is necessary to find ways to experimentally distinguish integrin mechano-signals from signals originating from cell-cell contacts and ion-channels. Also, the proposed interactions between certain ion-channels with integrins need to be characterized in more detail.

(iii) ROS: It would be important to clarify the mechanisms for how integrins regulate different NOXes and how they affect ROS release from mitochondria.

Although these are demanding tasks, every step towards a more detailed understanding of integrin signaling mechanisms and their interplay with other crucial molecules such as ROS, would bring us closer to understanding very important basic cellular functions and their roles in pathologies.

References

1. K. R. Legate, S. A. Wickstrom and R. Fassler, Genetic and cell biological analysis of integrin outside-in signaling, *Genes Dev.* **23**, 397–418 (2009).
2. A. Huttenlocher and A. R. Horwitz, Integrins in cell migration, *Cold Spring Harbor Perspectives in Biology* **3**, a005074 (2011).
3. S. Miyamoto, H. Teramoto, O. A. Coso *et al.*, Integrin function: Molecular hierarchies of cytoskeletal and signaling molecules, *J. Cell Biol.* **131**, 791–805 (1995).
4. S. W. Moore, P. Roca-Cusachs and M. P. Sheetz, Stretchy proteins on stretchy substrates: The important elements of integrin-mediated rigidity sensing, *Dev. Cell.* **19**, 194–206 (2010).
5. K. S. Zeller, A. Riaz, H. Sarve, J. Li, A. Tengholm and S. Johansson, The role of mechanical force and ROS in integrin-dependent signals, *PLoS One* **8**, e64897 (2013).
6. J. L. Walker and R. K. Assoian, Integrin-dependent signal transduction regulating cyclin D1 expression and G1 phase cell cycle progression, *Cancer Metastasis Rev.* **24**, 383–393 (2005).
7. R. K. Gupta and S. Johansson, Fibronectin assembly in the crypts of cytokinesis-blocked multilobular cells promotes anchorage-independent growth, *PLoS One* **8**, e72933 (2013).
8. J. Ivaska and J. Heino, Cooperation between integrins and growth factor receptors in signaling and endocytosis, *Annu. Rev. Cell. Dev. Biol.* **27**, 291–320 (2011).
9. K. S. Zeller, O. Idevall-Hagren, A. Stefansson *et al.*, PI3-kinase p110alpha mediates beta1 integrin-induced Akt activation and membrane protrusion during cell attachment and initial spreading, *Cell Signal.* **22**, 1838–1848 (2010).

10. C. Margadant and A. Sonnenberg, Integrin-TGF-beta crosstalk in fibrosis, cancer and wound healing, *EMBO Rep.* **11**, 97–105 (2010).

11. B. H. Luo, C. V. Carman and T. A. Springer, Structural basis of integrin regulation and signaling, *Annu. Rev. Immunol.* **25**, 619–647 (2007).

12. S. K. Akiyama, S. S. Yamada, K. M. Yamada and S. E. LaFlamme, Transmembrane signal transduction by integrin cytoplasmic domains expressed in single-subunit chimeras, *J. Biol. Chem.* **269**, 15961–15964 (1994).

13. V. P. Hytonen and V. Vogel, How force might activate talin's vinculin binding sites: SMD reveals a structural mechanism, *PLoS Comput. Biol.* **4**, e24 (2008).

14. Y. Sawada, M. Tamada, B. J. Dubin-Thaler *et al.*, Force sensing by mechanical extension of the Src family kinase substrate p130Cas, *Cell* **127**, 1015–1026 (2006).

15. A. J. Ehrlicher, F. Nakamura, J. H. Hartwig, D. A. Weitz and T. P. Stossel, Mechanical strain in actin networks regulates FilGAP and integrin binding to filamin A, *Nature* **478**, 260–263 (2011).

16. J. C. Friedland, M. H. Lee and D. Boettiger, Mechanically activated integrin switch controls alpha5beta1 function, *Science* **323**, 642–644 (2009).

17. D. E. Jaalouk and J. Lammerding, Mechanotransduction gone awry, *Nat. Rev. Mol. Cell Biol.* **10**, 63–73 (2009).

18. M. A. Wozniak and C. S. Chen, Mechanotransduction in development: A growing role for contractility, *Nat. Rev. Mol. Cell Biol.* **10**, 34–43 (2009).

19. A. J. Engler, S. Sen, H. L. Sweeney and D. E. Discher, Matrix elasticity directs stem cell lineage specification, *Cell* **126**, 677–689 (2006).

20. J. W. Shin and D. E. Discher, Cell culture: Soft gels select tumorigenic cells, *Nat. Mater.* **11**, 662–663 (2012).

21. R. Kaunas, P. Nguyen, S. Usami and S. Chien, Cooperative effects of Rho and mechanical stretch on stress fiber organization, *Proc. Natl. Acad. Sci. USA* **102**, 15895–15900 (2005).

22. M. H. Ali, P. T. Mungai and P. T. Schumacker, Stretch-induced phosphorylation of focal adhesion kinase in endothelial cells: Role of mitochondrial oxidants, *Am. J. Physiol. Lung. Cell Mol. Physiol.* **291**, L38–45 (2006).

23. R. Lutz, T. Sakai and M. Chiquet, Pericellular fibronectin is required for RhoA-dependent responses to cyclic strain in fibroblasts, *J. Cell. Sci.* **123**, 1511–1521 (2010).

24. S. Sukharev and F. Sachs, Molecular force transduction by ion channels: Diversity and unifying principles, *J. Cell. Sci.* **125**, 3075–3083 (2012).

25. Y. Sun, C. S. Chen and J. Fu, Forcing stem cells to behave: A biophysical perspective of the cellular microenvironment, *Annu. Rev. Biophys.* **41**, 519–542 (2012).

26. A. Arcangeli and A. Becchetti, Complex functional interaction between integrin receptors and ion channels, *Trends Cell Biol.* **16**, 631–639 (2006).

27. M. R. Mofrad, J. Golji, N. A. Abdul Rahim and R. D. Kamm, Force-induced unfolding of the focal adhesion targeting domain and the influence of paxillin binding, *Mech. Chem. Biosyst.* **1**, 253–265 (2004).

28. J. G. Wang, M. Miyazu, E. Matsushita, M. Sokabe and K. Naruse, Uniaxial cyclic stretch induces focal adhesion kinase (FAK) tyrosine phosphorylation followed by mitogen-activated protein kinase (MAPK) activation, *Biochem. Biophys. Res. Commun.* **288**, 356–361 (2001).

29. H. J. Hsu, C. F. Lee, A. Locke, S. Q. Vanderzyl and R. Kaunas, Stretch-induced stress fiber remodeling and the activations of JNK and ERK depend on mechanical strain rate, but not FAK, *PLoS One* **5**, e12470 (2010).

30. C. Guilluy, V. Swaminathan, R. Garcia-Mata *et al.*, The Rho GEFs LARG and GEF-H1 regulate the mechanical response to force on integrins, *Nat. Cell Biol.* **13**, 722–727 (2011).

31. Y. Sawada, K. Nakamura, K. Doi *et al.*, Rap1 is involved in cell stretching modulation of p38 but not ERK or JNK MAP kinase, *J. Cell. Sci.* **114**, 1221–1227 (2001).

32. H. B. Schiller, M. R. Hermann, J. Polleux *et al.*, beta1- and alphav-class integrins cooperate to regulate myosin II during rigidity sensing of fibronectin-based microenvironments, *Nat. Cell Biol.* **15**, 625–636 (2013).

33. O. Rossier, V. Octeau, J. B. Sibarita *et al.*, Integrins Beta1 and beta3 exhibit distinct dynamic nanoscale organizations inside focal adhesions, *Nat. Cell Biol.* **14**, 1057–1067 (2012).

34. R. K. Gupta and S. Johansson, $\beta 1$ integrins restrict the growth of foci and spheroids, *Histochem. Cell Biol.* **138**, 881–894 (2012).

35. K. Wennerberg, L. Lohikangas, D. Gullberg *et al.*, Beta 1 integrin-dependent and -independent polymerization of fibronectin, *J. Cell Biol.* **132**, 227–238 (1996).

36. K. Satoh, Y. Fukumoto and H. Shimokawa, Rho-kinase: Important new therapeutic target in cardiovascular diseases, *Am. J. Physiol. Heart Circ. Physiol.* **301**, H287–296 (2011).

37. P. Flevaris, A. Stojanovic, H. Gong, A. Chishti, E. Welch and X. Du, A molecular switch that controls cell spreading and retraction, *J. Cell Biol.* **179**, 553–565 (2007).

38. E. G. Arias-Salgado, S. Lizano, S. Sarkar *et al.*, Src kinase activation by direct interaction with the integrin beta cytoplasmic domain, *Proc. Natl. Acad. Sci. USA* **100**, 13298–13302 (2003).

39. E. H. Danen, P. Sonneveld, C. Brakebusch, R. Fassler and A. Sonnenberg, The fibronectin-binding integrins alpha5beta1 and alphavbeta3 differentially modulate RhoA-GTP loading, organization of cell matrix adhesions, and fibronectin fibrillogenesis, *J. Cell Biol.* **159**, 1071–1086 (2002).

40. N. P. Jones, J. Peak, S. Brader, S. A. Eccles and M. Katan, PLCgamma1 is essential for early events in integrin signalling required for cell motility, *J. Cell. Sci.* **118**, 2695–2706 (2005).

41. W. Droge, Free radicals in the physiological control of cell function, *Physiol. Rev.* **82**, 47–95 (2002).

42. V. Gogvadze, S. Orrenius and B. Zhivotovsky, Mitochondria in cancer cells: What is so special about them? *Trends Cell Biol.* **18**, 165–173 (2008).

43. M. Ushio-Fukai and Y. Nakamura, Reactive oxygen species and angiogenesis: NADPH oxidase as target for cancer therapy, *Cancer Lett.* **266**, 37–52 (2008).

44. D. P. Jones, Redox sensing: Orthogonal control in cell cycle and apoptosis signalling, *J. Int. Med.* **268**, 432–448 (2010).

45. N. K. Tonks, Redox redux: Revisiting PTPs and the control of cell signaling, *Cell* **121**, 667–670 (2005).

46. M. J. Morgan and Z. G. Liu, Crosstalk of reactive oxygen species and NF-kappaB signaling, *Cell Res.* **21**, 103–115 (2011).

47. E. Giannoni and P. Chiarugi, Redox Circuitries Driving Src Regulation, *Antioxid. Redox Signal.* (2013).

48. A. Aghajanian, E. S. Wittchen, S. L. Campbell and K. Burridge, Direct activation of RhoA by reactive oxygen species requires a redox-sensitive motif, *PLoS One* **4**, e8045 (2009).

49. L. Goitre, B. Pergolizzi, E. Ferro, L. Trabalzini and S. F. Retta, Molecular crosstalk between integrins and cadherins: Do reactive oxygen species set the talk? *J. Signal Transduct.* **2012**, 807682 (2012).

50. P. L. Hordijk, Regulation of NADPH oxidases: The role of Rac proteins, *Circ. Res.* **98**, 453–462 (2006).

51. A. S. Nimnual, L. J. Taylor and D. Bar-Sagi, Redox-dependent downregulation of Rho by Rac, *Nat. Cell Biol.* **5**, 236–241 (2003).

52. E. Werner and Z. Werb, Integrins engage mitochondrial function for signal transduction by a mechanism dependent on Rho GTPases, *J. Cell Biol.* **158**, 357–368 (2002).

53. G. Svineng, C. Ravuri, O. Rikardsen, N. E. Huseby and J. O. Winberg, The role of reactive oxygen species in integrin and matrix metalloproteinase expression and function, *Connect Tissue Res.* **49**, 197–202 (2008).

54. T. Fiaschi, G. Cozzi, G. Raugei *et al.*, Redox regulation of beta-actin during integrin-mediated cell adhesion, *J. Biol. Chem.* **281**, 22983–22991 (2006).

55. I. Lassing, F. Schmitzberger, M. Bjornstedt *et al.*, Molecular and structural basis for redox regulation of beta-actin, *J. Mol. Biol.* **370**, 331–348 (2007).

56. T. Fiaschi, G. Cozzi and P. Chiarugi, Redox regulation of nonmuscle myosin heavy chain during integrin engagement, *J. Signal Transduct.* **2012**, 754964 (2012).

57. J. S. Kim, T. Y. Huang and G. M. Bokoch, Reactive oxygen species regulate a slingshot-cofilin activation pathway, *Mol. Biol. Cell* **20**, 2650–2660 (2009).

58. M. Klemke, G. H. Wabnitz, F. Funke *et al.*, Oxidation of cofilin mediates T cell hyporesponsiveness under oxidative stress conditions, *Immunity* **29**, 404–413 (2008).

59. W. S. Wu, J. R. Wu and C. T. Hu, Signal cross talks for sustained MAPK activation and cell migration: The potential role of reactive oxygen species, *Cancer Metastasis Rev.* **27**, 303–314 (2008).

60. H. P. Monteiro, R. J. Arai and L. R. Travassos, Protein tyrosine phosphorylation and protein tyrosine nitration in redox signaling, *Antioxid. Redox Signal.* **10**, 843–889 (2008).

61. M. P. Murphy, How mitochondria produce reactive oxygen species, *Biochem. J.* **417**, 1–13 (2009).

62. M. Ushio-Fukai, Localizing NADPH oxidase-derived ROS, *Sci. STKE* **2006**, re8 (2006).

63. V. Jaquet, L. Scapozza, R. A. Clark, K. H. Krause and J. D. Lambeth, Small-molecule NOX inhibitors: ROS-generating NADPH oxidases as therapeutic targets, *Antioxid. Redox Signal.* **11**, 2535–2552 (2009).

64. J. D. Lambeth, T. Kawahara and B. Diebold, Regulation of Nox and Duox enzymatic activity and expression, *Free Radic. Biol. Med.* **43**, 319–331 (2007).

65. B. C. Dickinson and C. J. Chang, Chemistry and biology of reactive oxygen species in signaling or stress responses, *Nat. Chem. Biol.* **7**, 504–511 (2011).

66. M. L. Taddei, M. Parri, T. Mello *et al.*, Integrin-mediated cell adhesion and spreading engage different sources of reactive oxygen species, *Antioxid. Redox Signal.* **9**, 469–481 (2007).

67. P. Chiarugi, G. Pani, E. Giannoni *et al.*, Reactive oxygen species as essential mediators of cell adhesion: The oxidative inhibition of a FAK tyrosine phosphatase is required for cell adhesion, *J. Cell Biol.* **161**, 933–944 (2003).

68. K. Dib, F. Melander, L. Axelsson *et al.*, Down-regulation of Rac activity during beta 2 integrin-mediated adhesion of human neutrophils, *J. Biol. Chem.* **278**, 24181–24188 (2003).

69. F. Buricchi, E. Giannoni, G. Grimaldi *et al.*, Redox regulation of ephrin/integrin cross-talk, *Cell Adh. Migr.* **1**, 33–42 (2007).

70. G. W. De Keulenaer, D. C. Chappell, N. Ishizaka *et al.*, Oscillatory and steady laminar shear stress differentially affect human endothelial redox state: Role of a superoxide-producing NADH oxidase, *Circ. Res.* **82**, 1094–1101 (1998).

71. Y. Zhang, F. Peng, B. Gao, A. J. Ingram and J. C. Krepinsky, Mechanical strain-induced RhoA activation requires NADPH oxidase-mediated ROS generation in caveolae, *Antioxid. Redox Signal.* **13**, 959–973 (2010).

72. S. B. Lee, I. H. Bae, Y. S. Bae and H. D. Um, Link between mitochondria and NADPH oxidase 1 isozyme for the sustained production of reactive oxygen species and cell death, *J. Biol. Chem.* **281**, 36228–36235 (2006).

73. M. Karlsson, T. Kurz, U. T. Brunk, S. E. Nilsson and C. I. Frennesson, What does the commonly used DCF test for oxidative stress really show? *Biochem. J.* **428**, 183–190 (2010).

74. A. J. Meyer and T. P. Dick, Fluorescent protein-based redox probes, *Antioxid. Redox Signal.* **13**, 621–650 (2010).

75. M. P. Murphy, A. Holmgren, N. G. Larsson *et al.*, Unraveling the biological roles of reactive oxygen species, *Cell Metab.* **13**, 361–366 (2011).

76. S. A. Santi and H. Lee, The Akt isoforms are present at distinct subcellular locations, *Am. J. Physiol. Cell Physiol.* **298**, C580–591 (2010).

77. A. Riaz, K. S. Zeller and S. Johansson, Receptor-specific mechanisms regulate phosphorylation of AKT at Ser473: Role of RICTOR in beta1 integrin-mediated cell survival, *PLoS One* **7**, e32081 (2012).

78. N. Wang, J. D. Tytell and D. E. Ingber, Mechanotransduction at a distance: Mechanically coupling the extracellular matrix with the nucleus, *Nat. Rev. Mol. Cell Biol.* **10**, 75–82 (2009).

79. L. L. Hilenski, R. E. Clempus, M. T. Quinn, J. D. Lambeth and K. K. Griendling, Distinct subcellular localizations of Nox1 and Nox4 in vascular smooth muscle cells, *Arterioscler. Thromb. Vasc. Biol.* **24**, 677–683 (2004).

80. K. Chen, M. T. Kirber, H. Xiao, Y. Yang and J. F. Keaney Jr., Regulation of ROS signal transduction by NADPH oxidase 4 localization, *J. Cell Biol.* **181**, 1129–1139 (2008).

81. K. D. Martyn, L. M. Frederick, K. von Loehneysen, M. C. Dinauer and U. G. Knaus, Functional analysis of Nox4 reveals unique characteristics compared to other NADPH oxidases, *Cell Signal.* **18**, 69–82 (2006).

82. B. C. Dickinson, C. Huynh and C. J. Chang, A palette of fluorescent probes with varying emission colors for imaging hydrogen peroxide signaling in living cells, *J. Am. Chem. Soc.* **132**, 5906–5915 (2010).

83. B. C. Dickinson, J. Peltier, D. Stone, D. V. Schaffer and C. J. Chang, Nox2 redox signaling maintains essential cell populations in the brain, *Nat. Chem. Biol.* **7**, 106–112 (2011).

84. M. Malinouski, Y. Zhou, V. V. Belousov, D. L. Hatfield and V. N. Gladyshev, Hydrogen peroxide probes directed to different cellular compartments, *PLoS One* **6**, e14564 (2011).

85. M. B. Cannon and S. J. Remington, Re-engineering redox-sensitive green fluorescent protein for improved response rate, *Protein Sci.* **15**, 45–57 (2006).

86. G. T. Hanson, R. Aggeler, D. Oglesbee *et al.*, Investigating mitochondrial redox potential with redox-sensitive green fluorescent protein indicators, *J. Biol. Chem.* **279**, 13044–13053 (2004).

87. K. A. Lukyanov and V. V. Belousov, Genetically encoded fluorescent redox sensors, *Biochim. Biophys. Acta* **1840**(2), 745–756 (2014).

Chapter 9

Can Mathematical Models Predict the Outcomes of Prostate Cancer Patients Undergoing Intermittent Androgen Deprivation Therapy?

R. A. Everett[*], A. M. Packer[*] and Y. Kuang[*,†]

[*]School of Mathematical and Statistical Sciences
Arizona State University, Tempe, AZ 85287, USA

[†]Department of Mathematics
King Abdulaziz University
Jeddah 21589, Saudi Arabia

Androgen deprivation therapy is a common treatment for advanced or metastatic prostate cancer. Like the normal prostate, most tumors depend on androgens for proliferation and survival but often develop treatment resistance. Hormonal treatment causes many undesirable side effects which significantly decrease the quality of life for patients. Intermittently applying androgen deprivation in cycles reduces the total duration with these negative effects and may reduce selective pressure for resistance. We extend an existing model which used measurements of patient testosterone levels to accurately fit measured serum prostate specific antigen (PSA) levels. We test the model's predictive accuracy, using only a subset of the data to find parameter values. The results are compared with those of an existing piecewise linear model which does not use testosterone as an input. Since actual treatment protocol is to re-apply therapy when PSA levels recover beyond some threshold value, we develop a second method for predicting the PSA levels. Based on a small set of data from seven patients, our results showed that the piecewise linear model produced slightly more accurate results while the two predictive methods are comparable. This suggests that a simpler model may be more beneficial for a predictive use compared to a more biologically insightful model, although further research is needed in this field prior to implementing mathematical models as a predictive method in a clinical setting. Nevertheless, both models are an important step in this direction.

1. Introduction

1.1. Prostate cancer and treatment

The probability of an American man developing prostate cancer in a lifetime is 1 in 6.[28] Although the incidence and death trends for prostate cancer are declining, there is still no curative treatment for patients with distant metastases.[7] Androgen deprivation therapy (ADT) is one of the most common and effective therapies for patients with metastatic cancer[22] and has recently been used to also treat non-metastatic disease.[4,17] Although the initial response rate of ADT is above 90%, most patients become resistant to treatment and develop castration-resistant prostate cancer (CRPC).[22] CRPC is usually fatal with a median survival time of 2.5 to 3 years.[22,12]

Androgens, specifically testosterone and 5α-dihydrotestosterone (DHT), are essential for maintenance of the prostate. Prostate secretory epithelial cells depend on androgens for proliferation and survival. The testes produce 90–95% of the androgens in the body with the adrenal gland producing the remainder.[7] Androgens regulate cellular proliferation and survival via activation of the androgen receptor (AR), a nuclear hormone receptor. Around 90% of serum testosterone that enters the prostate is enzymatically converted to DHT, which has a greater affinity for AR than that of testosterone. Ligand binding to AR causes a cascade of events that upregulate proliferation, survival, and secretion of prostate specific antigen (PSA).[6] Serum PSA is used as a biomarker for prostate cancer because PSA expression is maintained by cancerous cells. While its effectiveness as a diagnostic tool is controversial, PSA is useful for gauging the response of disease to ADT. ADT inhibits AR signaling by blocking androgen production and AR binding. Therapy induces regression of both mass and PSA secretion by the prostate and cancer.

ADT can be performed by surgical or chemical castration. Orchiectomy, the removal of the testes, is a relatively simple procedure that results in a decrease of testosterone levels. However, chemical castration is more common due to the psychological effects of the surgery.[27,18] Current chemical castration options include luteinizing hormone release hormone (LHRH) agonists, Gonadotropin releasing hormone (GnRH) antagonists, and anti-androgens. A combination of an anti-androgen and a LHRH agonist is called total androgen blockade.[6,7]

Intermittent androgen deprivation (IAD) therapy consists of alternating periods of on- and off-treatment and provides many benefits over continuous (CAD) therapy, including increased health related quality of life, reduced therapy costs (LHRH agonists cost about $300 to $400 a month[17]), and potentially delaying resistance to treatment, although the latter remains controversial.[24,20,17,8] ADT causes numerous side effects such as erectile dysfunction, loss of libido, gynecomastia, osteoporosis, and anemia.[9,17] Some of these side effects can potentially lead to more serious conditions, such as diabetes, hypertension, and cardiovascular disease.[9] Two recent studies by Crook *et al.*[4] and Hussain *et al.*[12] compared IAD to CAD therapy in patients with prostate cancer. Crook *et al.* concluded that IAD was noninferior to CAD in terms of survival, but improvements in quality of life were observed in IAD patients.[4] Hussain *et al.* observed small improvements in quality of life for IAD patients and but their findings in terms of survival were statistically insignificant; they were not able to rule out a greater risk of death from IAD compared to CAD nor rule out significant inferiority of IAD.[12] Crook *et al.* and Hussian *et al.* provide two examples of recent studies to determine if IAD or CAD is more effective at delaying resistance to treatment. Although this topic remains controversial,[24,20,17,8] the European Association of Urology (EAU) claims IAD as the standard of care for patients with metastatic or biochemically recurrent prostate cancer.[20]

1.2. *Recent works*

While IAD offers several benefits, there are still controversies in how the treatment should be applied, such as who should receive IAD therapy, when to start and stop therapy, and what thresholds should be used for starting and stopping treatment.[17,26] Mathematical models are important tools for achieving improved therapy and determining the patient-specific answers to some of these controversies. In 2004, Jackson[15,16] used a system of partial differential equations to investigate the mechanisms for CRPC. The model assumed the tumor comprised of two types of cells, androgen-dependent (AD) and androgen-independent (AI), with the latter contributing to the AI tumor relapse and resistance to therapy. The proliferation and death rates of both cell types differed in various androgen environments; during androgen deprivation, the AD proliferation rate decreased and the AD death rate increased while the AI proliferation rate remained constant and the AI death rate decreased. The results agreed with experimental data, capturing the exponential growth pre-treatment, androgen-sensitivity following therapy, and tumor regrowth.

Ideta *et al.*[13] presented an ordinary differential equation model consisting of AI and AD cell populations in order to compare CAD and IAD with respect to relapses. They also assumed a decrease in AD proliferation rate and increase in AD death rate during androgen deprivation, however they considered three possible net growth rates for AI cells. Their model included mutations from AD to AI cells.

Hirata *et al.*[11,10] considered three cell populations: AD, reversible AI, and irreversible AI. The reversible AI cells, possibly created by adaptations, can revert back to AD cells, whereas the irreversible AI cells cannot. Similarly to Ideta*et al.*, the irreversible changes can be due to mutations. The model was fit to clinical data and used to group patients into categories based on the IAD versus CAD and the prevention of relapse: IAD may prevent a relapse, IAD may delay a relapse, and CAD is more effective in delaying a relapse than IAD. In another investigation, the first two and one-half cycles of treatment were used to find individualized parameters and then predict PSA responses to subsequent treatment. This approach was presented as a basis for future methods of individualized cancer treatment.

Portz *et al.*[23] developed a novel model of ADT by extending mathematical frameworks in ecology to the two-subpopulation models of ADT.[16,13] The cell quota model,[5] which relates growth to an intracellular nutrient, was used for proliferation of both the AD and AI cell populations. Since AR signaling reflects the intracellular androgen-AR interactions, the "cell quota" was conceived as intracellular androgen concentrations. A significant difference in this model from previous work[13] was that the so-called AI cells were assumed be responsive to androgens and that PSA production was androgen-independent. The bidirectional mutation rates and cell-specific rates of PSA production were also functions of the cell quota. Cells had a constant death rate and also produced PSA at a constant, baseline rate. The model was validated with clinical data[1] and its accuracy compared to that of the Ideta *et al.*[13] model. The androgen quota model exhibited significantly greater accuracy for each patient data set. Their results supported the idea that ADT models should assume that AI cells maintain sensitivity to androgens, though to a lesser degree than AD cells. The model was also used to predict future hypothetical treatment cycles. However, unlike the method used in Ref. 11, predictive accuracy was not assessed using subsets of the data. While their conclusions provided information about the mechanisms of resistance, their patient specific predictions lack validity.

1.3. Methods and findings

A mathematical model that accurately predicts the next cycle of treatment for an individual patient undergoing IAD therapy is an important tool that can potentially be used in a clinical setting. Here, we first extend the model from Portz et al.[23] (Model 1) by adding an androgen-dependent cell death rate. This extended model has been validated with the same clinical data.[21] Different methods (Methods 1, 2) are then implemented for measuring the accuracy of the model's predictions when using an increasing subset of data. Parameters are found for the first treatment cycle, then used to predict the observed response to the second cycle, and so forth. The results are compared to those obtained using the model by Hirata et al.[11] (Model 2) to make predictions of the same data. Finally, both models are used to predict patient response to a hypothetical future treatment cycle. The predictions produced by Model 1 Method 1 were not very accurate. Model 1 Method 2 and Model 2 Method 2 produced more accurate predictions, although the timing of the predictions was often incorrect. Model 2 Method 1 was also more accurate, but often either had incorrect timing or under-predicted the results. Our results suggest that a simpler model may be more beneficial for a predictive use and that further research is needed in this field prior to implementing mathematical models as a predictive method in a clinical setting.

2. Mathematical Models

2.1. Model 1: Extension of model by Portz et al.

We propose the following prostate cancer model,[21] which is an extension of the model by Portz et al.[23] with death rates dependent on cell androgen quotas:

$$\frac{dX_1}{dt} = \underbrace{\mu_m \left(1 - \frac{q_1}{Q_1}\right) X_1}_{\text{proliferation}} - \underbrace{D_1(Q_1)X_1}_{\text{death}} \underbrace{- \lambda_1(Q_1)X_1 + \lambda_2(Q_2)X_2}_{\text{switching}} \tag{1}$$

$$\frac{dX_2}{dt} = \underbrace{\mu_m \left(1 - \frac{q_2}{Q_2}\right) X_2}_{\text{proliferation}} - \underbrace{D_2(Q_2)X_2}_{\text{death}} \underbrace{- \lambda_2(Q_2)X_2 + \lambda_1(Q_1)X_1}_{\text{switching}} \tag{2}$$

$$\frac{dQ_i}{dt} = \underbrace{v_m \frac{q_m - Q_i}{q_m - q_i} \frac{A}{A + v_h}}_{\text{uptake}} - \underbrace{\mu_m(Q_i - q_i)}_{\text{dilution}} - \underbrace{bQ_i}_{\text{degradation}} \quad (3)$$

$$\frac{dP}{dt} = \underbrace{\sigma_0(X_1 + X_2)}_{\text{baseline production}} + \underbrace{\sigma_1 X_1 \frac{Q_1^m}{Q_1^m + \rho_1^m} + \sigma_2 X_2 \frac{Q_2^m}{Q_2^m + \rho_2^m}}_{\text{androgen-dependent production}} - \underbrace{\varepsilon P}_{\text{clearance}} \quad (4)$$

where

$$D_i(Q_i) = \underbrace{d_i \frac{R_i^\alpha}{Q_i^\alpha + R_i^\alpha}}_{\text{AD Apoptosis}} + \underbrace{\delta_i}_{\text{AI Death}}$$

and

$$\underbrace{\lambda_1(Q) = c_1 \frac{K_1^n}{Q^n + K_1^n}}_{\text{CS to CR}}, \quad \underbrace{\lambda_2(Q) = c_2 \frac{Q^n}{Q^n + K_2^n}}_{\text{CR to CS}}$$

X_1, X_2 represent the AD and AI cell populations, respectively. The terms "androgen-dependent" and "androgen-independent" have been used previously in both mathematical models as well as in biological literature.[14,6,15,13,10] However, "androgen-independent" cells are often not completely independent, but have a lower threshold for androgens. Thus, we refer to AD and AI cells as "castration-sensitive" (CS) and "castration-resistant" (CR), respectively, as seen in recent literature.[25,18,22,7] The proliferation rates are given by Droop's model, which is dependent upon some cell quota or limiting nutrient. Here, the cell quota (Q) is intracellular androgen. μ_m represents the maximum proliferation rate and q_i is the minimum cell quota. Since CR cells are able to proliferate at lower levels of androgen, $q_2 < q_1$.

Portz *et al.*[23] assume the cell death rate is constant for simplicity. Our extension of the model is the incorporation of an androgen-dependent death rate in addition to the constant death rate δ_i. d_i represents the maximum androgen-dependent death rate. The shape parameters R_i and α represent the half saturation level and hill coefficient, respectively, which describe the cell death rate sensitivity to the cell quota level. Whereas Jackson[16,15] and Ideta *et al.*[13] assume the AI death rate decreases as the androgen concentration decreases, we assume the death rates increases as the androgen concentration decreases, which is supported by biological results.[6,25]

The model also assumes androgen-dependent mutation rates, λ_i, to account for switching between the cell populations. c_1 and c_2 represent the maximum switching rates. K_i and n represent the half saturation level and hill coefficient, respectively, which describe the cell switching sensitivity to the cell quota level. We interpret these switching rates as both accommodative and adaptive switching.[21]

As serum androgen A increases, the cell uptakes more androgen and approaches the maximum. This maximum uptake rate is regulated by the cell quota $Q(t)$, maximum cell quota q_m, minimum cell quota q_i, and maximum uptake rate v_m. v_h represents the uptake half saturation level. $\mu_m(Q_i - q_i)$ represents the amount of cell quota used in the cell for growth. The cell quota degrades at rate b.

PSA, P, is produced at both a baseline rate σ_0 and androgen-dependent rate by both cell populations. $\sigma_{1,2}$ represent the maximum androgen dependent PSA productions by the two cell populations. The shape parameters ρ_i and m represent the half saturation level and hill coefficient, respectively, which describe the PSA production rate sensitivity to the cell quota level. PSA is cleared from the blood at rate ε. For further details on the Portz *et al.* model formulation and explanation, see Ref. 23.

2.2. *Model 2: Model by Hirata et al.*

We compare the predictions produced using Model 1 to the predictions produced from a model by Hirata *et al.*[10,11] This model considered an AD cell population (x_1), a reversible AI cell population (x_2), and an irreversible AI cell population (x_3), modeled by the following:

$$\frac{d}{dt}\begin{pmatrix} x_1(t) \\ x_2(t) \\ x_3(t) \end{pmatrix} = \begin{pmatrix} w_{1,1}^1 & 0 & 0 \\ w_{2,1}^1 & w_{2,2}^1 & 0 \\ w_{3,1}^1 & w_{3,2}^1 & w_{3,3}^1 \end{pmatrix} \begin{pmatrix} x_1(t) \\ x_2(t) \\ x_3(t) \end{pmatrix} \tag{5}$$

for the on-treatment periods and

$$\frac{d}{dt}\begin{pmatrix} x_1(t) \\ x_2(t) \\ x_3(t) \end{pmatrix} = \begin{pmatrix} w_{1,1}^0 & w_{1,2}^0 & 0 \\ 0 & w_{2,2}^0 & 0 \\ 0 & 0 & w_{3,3}^0 \end{pmatrix} \begin{pmatrix} x_1(t) \\ x_2(t) \\ x_3(t) \end{pmatrix} \tag{6}$$

for the off-treatment periods. The PSA levels P are modeled by the following:

$$P = x_1 + x_2 + x_3 \tag{7}$$

Whereas Model 1 captures the intermittent property using serum androgen levels as an input, Model 2 uses a binary on- or off-treatment input. Following Hirata *et al.*, the parameters were constrained so that the nondiagonal parameters are non-negative, $w_{3,3}^1 \geq 0$, and the cell class can change its volume by at most 20% per day, namely $|\sum_{i\in\{1,2,3\}} w_{i,j}^m| < 0.2$, where $j \in \{1,2,3\}$ and $m \in \{0,1\}$. See Refs. 10 and 11 and for further model details.

3. Data and Simulations

Akakura *et al.*[1] published results from a study with seven patients undergoing intermittent androgen deprivation therapy. Four of the men (patients 1, 2, 3, 5) had stage C cancer, in which the cancer has spread outside the prostate, but not yet to other parts of the body; one man (patient 4) had stage D1 cancer, in which the cancer has only spread to local lymph nodes; two men (patients 6, 7) had stage D2 or metastasized cancer.[2] The data consisted of serum PSA and testosterone levels, obtained at monthly intervals. Patients received goserelin acetate (LHRH agonist) and cyproterone acetate (anti-androgen) until the PSA level reached a normal level and remained in this range for about four months, although this timing varied greatly among the patients. The patient then stayed off therapy until the levels reached about 20 ng/mL. It should be noted that Akakura *et al.* state that the upper limit of 20 ng/mL was set arbitrarily and also seems to vary among the patients. For more information on the study, see Ref. 1.

Since we use the PSA data to verify the models, we are able to use the androgen data directly for Model 1. Following Portz *et al.*,[23] we interpolated the data using piecewise cubic hermit splines and an exponential function between the last off-treatment $A(t_i)$ and first on-treatment $A(t_f)$ data points with $l = 1$:

$$A(t) = A(t_f) + (A(t_i) - A(t_f))e^{-(\gamma/l)(t-t_i)}. \tag{8}$$

This equation was also used for the predicted off-treatment PSA growth with $l = 100$ for Method 2 (Sec. 3.2).

After first fitting the free parameters by hand, we used the Nelder-Mead simplex algorithm[19] to find the free parameters that minimized the mean square error (MSE) between the PSA data and model. The fixed parameter values as well as the free parameter ranges for Model 1 and Model 2 can be found in Tables 1 and 2 respectively. In order to test the accuracy of the prediction, we first find the parameters using only 1.5 cycles of data. Using these parameters, we then run the model for another treatment cycle and compute the error between the future model and the remaining, or "future", data. We repeat this process using 2.5 cycles of data and then 3.5 cycles of data when possible.

In order to make future PSA predictions, we must first generate future serum androgen levels. We propose two different methods, described below, for generating these future androgen levels and then apply these methods to Model 2. To compare these methods and models, we compute both the MSE and mean relative error (MRE) (Table 5) as well as plot the results. The figures compare Model 1 and Model 2, each using both prediction methods, to clinical data for both the PSA levels (ng/ML) and the serum androgen levels (nM) where applicable. The right of the vertical dashed line represents the prediction with the "future" data overlaid for comparison.

3.1. *Prediction method* 1: *Average function*

We implemented the method used by Portz *et al.*[23] for generating future serum androgen levels, which consists of generating a rectangular function based on the average off- and on-treatment serum androgen values and off- and on-treatment durations. To apply this method to Model 2, we set the on- and off-treatment binary switch to occur after mean durations of on- and off-treatment.

Table 1.   Model 1 parameter ranges.

Para.	Meaning	Value	Reference
μ_m	Maximum proliferation rate	0.009–0.045/day	3
q_1	Minimum CS cell quota	0.19–0.29 nM	23
q_2	Minimum CR cell quota	0.10–0.21 nM	23
σ_1	CS PSA production rate	0.0001–0.28 nag/mL/cell/day	
σ_2	CR PSA production rate	0.06–0.36 ng/mL/cell/day	
σ_0	Baseline PSA production rate	0–0.031 ng/mL/cell/day	
d_1	Maximum CS CDR	0.0035–0.029 day^{-1}	*
d_2	Maximum CR CDR	0.0019–0.0059 day^{-1}	*
R_1	CS CDR half-saturation level	0.46–3.02 nM	*
R_2	CR CDR half-saturation level	0.96–6.17 nM	*
δ_1	CS androgen independent death rate	0.0006–0.0083 day^{-1}	*
δ_2	CR androgen independent death rate	0.011–0.042 day^{-1}	*
c_1	Maximum CS to CR mutation rate	0.00016 day^{-1}	13, 23
c_2	Maximum CR to CS mutation rate	0.00012 day^{-1}	23
K_1	CS to CR mutation half-saturation level	0.8 nM	23
K_2	CR to CS mutation half-saturation level	1.7 nM	23
n	Selection function exponent	3	
q_m	Maximum cell quota	5 nM	23
v_m	Maximum uptake rate	0.27 nM/day	23
v_h	Uptake rate half-saturation level	4 nM	23
b	Intracellular androgen degradation rate	0.09 day^{-1}	23
ρ_1	CS PSA production half-saturation level	1.3 nM	23
ρ_2	CR PSA production half-saturation level	1.1 nM	23
m	PSA production function exponent	3	23
ε	PSA clearance rate	0.08 day^{-1}	23
α	CDR function exponent	3	

*Indicates values such that total cell death rate (CDR) is within biological ranges.[3,13]

Table 2.   Model 2 parameter ranges. rAI represents reversible AI cells and irrAI represents irreversible AI cells.

Para.	Meaning	Value	Reference
w_{11}^1	on-treat. AD growth rate	−0.15–−0.015	11, 10
w_{22}^1	on-treat. rAI growth rate	−0.015–0.0009	11, 10
w_{33}^1	on-treat. irrAI growth rate	0.002–0.003	11, 10
w_{21}^1	on-treat. AD to rAI influx rate	0.0006–0.002	11, 10
w_{31}^1	on-treat. AD to irrAI influx rate	0.0003–0.001	11, 10
w_{32}^1	on-treat. rAI to irrAI influx rate	0–0	11, 10
w_{11}^0	off-treat. AD growth rate	0.001–0.003	11, 10
w_{22}^0	off-treat. rAI growth rate	0.002–0.008	11, 10
w_{33}^0	off-treat. irrAI growth rate	−0.13–−0.0044	11, 10
w_{12}^0	off-treat. rAI to AD influx rate	0.049–0.18	11, 10

3.2. *Prediction method 2: Threshold function*

During the clinical trial,[1] the treatment resumed once the PSA levels reached the approximate threshold of 20 ng/mL. This implies that the future androgen levels should depend on the future PSA level. In this method, once the mean on-treatment duration occurs, the androgen level increases using Eq. (8) until a PSA threshold is reached and then decays according to Eq. (8). Similarly, for Model 2, the treatment remained off until the PSA levels reached a threshold and then the model switched to the on-treatment equations.

4. Results

Portz *et al.*[23] used all the provided data to predict the following cycle. However, in doing so, they were not able to test the accuracy of their predictions. We first use an extension of their model (Model 1) and their method (Method 1) and determine the accuracy of the predictions. We then repeat the process three more times to compare the predictions produced by the two models each using the two methods. These results are summarized in Table 3. A description of the patient-specific predictions are found in Table 4 and the patient-specific errors are found in Table 5. In the following subsections, we discuss the patient 1 predictions and the general results for each model and method.

4.1. *Model 1 Method 1*

The model is extremely accurate when fitting the used data, however the model is not always accurate when predicting the future cycle (Table 5). When using

Table 3. Method and model prediction comparison summary. Under refers to under-predicting the PSA levels.

	Model 1	Model 2
Method 1	Not accurate	More accurate, incorrect timing or under
Method 2	More accurate, incorrect timing	More accurate, incorrect timing

Table 4. Testable patient-specific prediction summary. Under refers to under-predicts, over refers to over-predicts, and shift refers to a phase-shift. Cycle refers to the number of cycles of data used to make the prediction.

Patient	Cycle	Description			
		Model 1		Model 2	
		Meth. 1	Meth. 2	Meth. 1	Meth. 2
1	1.5	under	shift	mostly accurate	mostly accurate
1	2.5	over	shift	shift	shift
2	1.5	under, shift	shift	under, shift	shift
3	1.5	over, shift	shift	shift	shift
4	1.5	under	under, shift	under	over, shift

Table 5.   Prediction errors. Cycle refers to the number of cycles of data used to make the prediction.

Patient	Cycle	Future MSE				Future MRE			
		Model 1		Model 2		Model 1		Model 2	
		Meth. 1	Meth. 2	Meth. 1	Meth. 2	Meth. 1	Meth. 2	Meth. 1	Meth. 2
1	1.5	20.26	34.50	17.02	17.05	.7190	2.076	.9650	.9665
1	2.5	111.7	42.23	22.41	22.62	2.433	1.278	.7278	.7519
2	1.5	27.80	47.83	22.67	50.628	.6249	.9888	.7510	1.0591
3	1.5	319.0	213.46	224.98	206.5	.5176	.2928	.4726	.4586
4	1.5	45.54	54.40	19.43	54.21	.5125	.5517	.3853	.7026

1.5 cycles of data for patient 1 (Fig. 1), the model under-predicts the PSA levels, only reaching about 6 ng/mL, when in reality the patient's levels reached about 13 ng/mL. In a clinical setting, the model would indicate that the patient could continue off-treatment when in reality the patient resumed treatment. When assuming 2.5 cycles of data, the model over-predicts the PSA levels, reaching about 37 ng/mL, when in reality the patient reached levels around 20 ng/mL before resuming treatment. In this case, the model would suggest the patient resume treatment much sooner, shortening their off-treatment period. With all 3.5 cycles of data, the model again suggests high PSA levels, however we are not able to test the accuracy of the fourth cycle due to a limited amount of data. Similarly for patients 2–4 (Figs. 2, 3, 4), the predictions are not very accurate in predicting the PSA levels for various reasons (Tables 4, 5). Since the data for patients 5–7 (Fig. 5) consisted of only 1.5 cycles of data, we were not able to test the accuracy of the predictions.

4.2. Model 1 Method 2

We repeated the process of predicting the outcomes of patients using Model 1 with Method 2. For patient 1 (Fig. 1), assuming 1.5 cycles of data, Method 2 much more accurately predicts the maximum PSA level (about 13 ng/mL) compared to Method 1, although it takes more days to reach this maximum compared to the data. This "shift" in the PSA levels explains the high error values (Table 5). Similarly, when using 2.5 cycles of data, the predicted PSA levels are "shifted" compared to the data and thus begin the fourth cycle early. However, the MSE and MRE values are smaller than with Method 1. When using all 3.5 cycles of data, Method 2 predicts a maximum PSA level similar to that of the previous cycles, which is much smaller than the predicted PSA levels using Method 1. In general, Method 2 more accurately predicts the PSA peak, as expected, but the timing is often incorrect as seen by the "shift" in PSA levels.

4.3. Model 2 Method 1

We repeated the process a third time using Model 2 with Method 1. For patient 1 using 1.5 cycles of data (Fig. 1), the model accurately predicts the PSA level

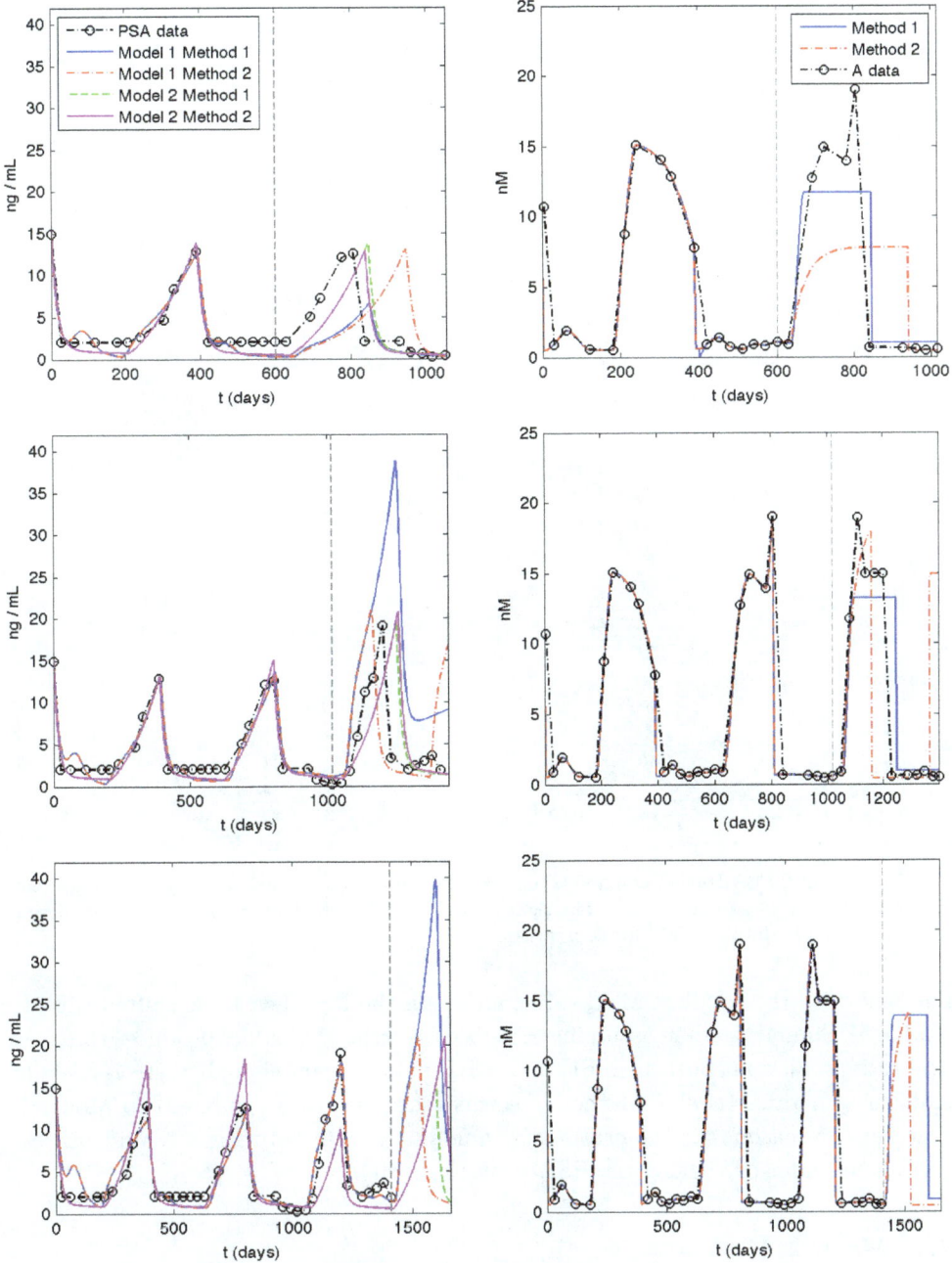

Fig. 1. Patient 1 PSA levels (left) and serum androgen levels (right) using 1.5 cycles of data (top row), 2.5 cycles (second row), and all 3.5 cycles (third row). The right of the vertical dashed line represents the prediction with the "future" data overlaid for comparison.

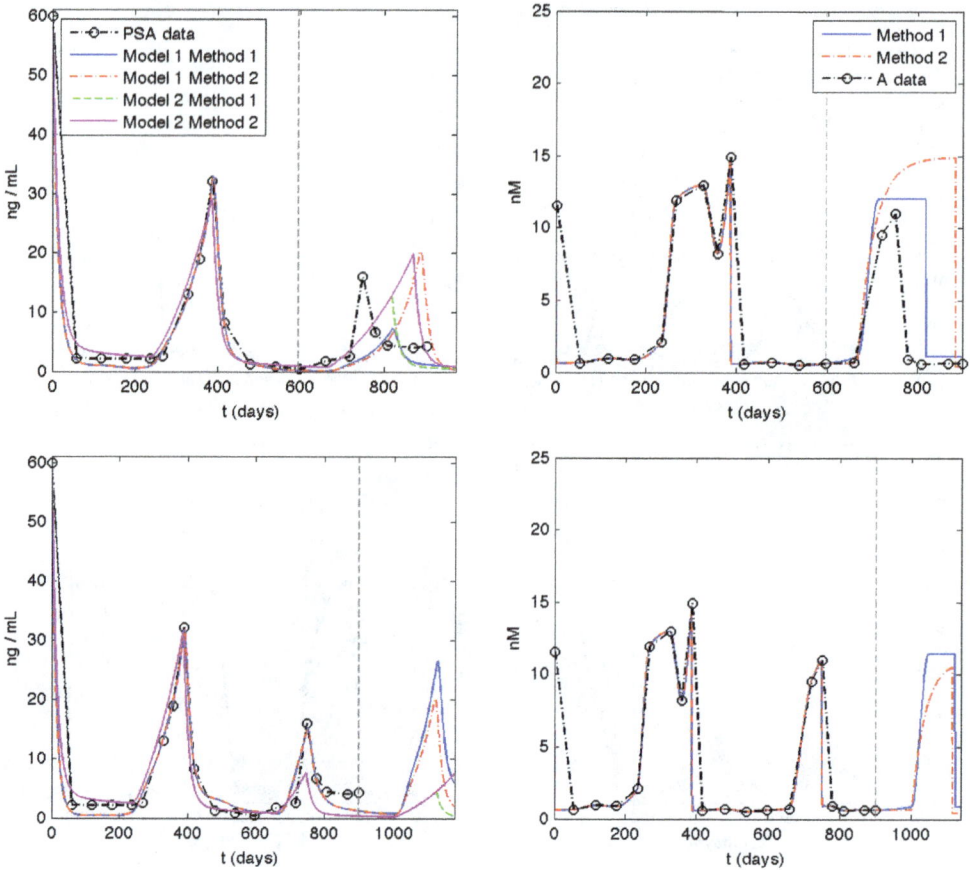

Fig. 2.   Patient 2 PSA levels (left) and serum androgen levels (right) using 1.5 cycles of data (top row) and all 2.5 cycles (second row). The right of the vertical dashed line represents the prediction with the "future" data overlaid for comparison.

outcome with the smallest MSE value, although the PSA levels are shifted slightly (Table 5). Similarly, when assuming 2.5 cycles of data, the model produces the most accurate prediction, both in MSE and MRE values, even though there is a slight shift. In general, Model 2 Method 1 seems more accurate compared to Model 1, however the reasons for the errors vary among patients between a "shift" in PSA levels and under-predicting the PSA peak (Table 4).

4.4. Model 2 Method 2

Model 2 with Method 2 produced similar results to Model 2 with Method 1. Both predictions increase at the same rate, however the timing for the switch to on-treatment is different, by design of the methods. For patient 1 (Fig. 1) assuming both 1.5 and 2.5 cycles of data, Methods 1 and 2 produce very similar results. When using all 3.5 cycles, Method 2 switches to on-treatment after Method 1, producing

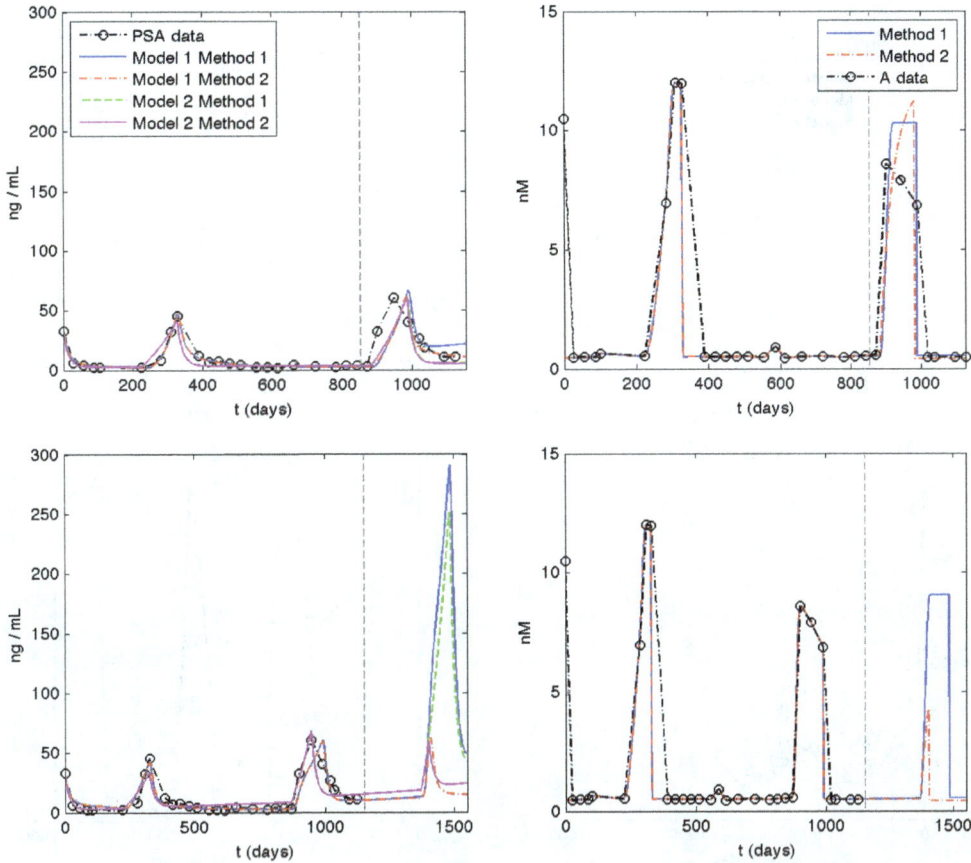

Fig. 3.   Patient 3 PSA levels (left) and serum androgen levels (right) using 1.5 cycles of data (top row) and all 2.5 cycles (second row). The right of the vertical dashed line represents the prediction with the "future" data overlaid for comparison.

a larger maximum PSA value. When using 1.5 cycles, Model 2 produces a higher rate of increase in PSA levels than Model 1. However when assuming 2.5 and 3.5 cycles, Model 1 produces higher rate of increase in PSA levels. In general, similarly to Model 1 Method 2, Model 2 Method 2 predicts the peak PSA values well, but the timing is often incorrect (Tables 3, 4).

5. Discussion

We extend the work of Portz et al.[23] by first modifying the model to be biologically more accurate and testing the accuracy of the predictions. Similarly to Hirata et al.,[11,10] we use a portion of the data to find the best patient specific parameters and then use these parameters to predict the next cycle of treatment. To determine the accuracy of the prediction, we calculate the MSE and MRE values for the predicted cycle, i.e. to the right of the vertical dashed gray line (Table 5). We then

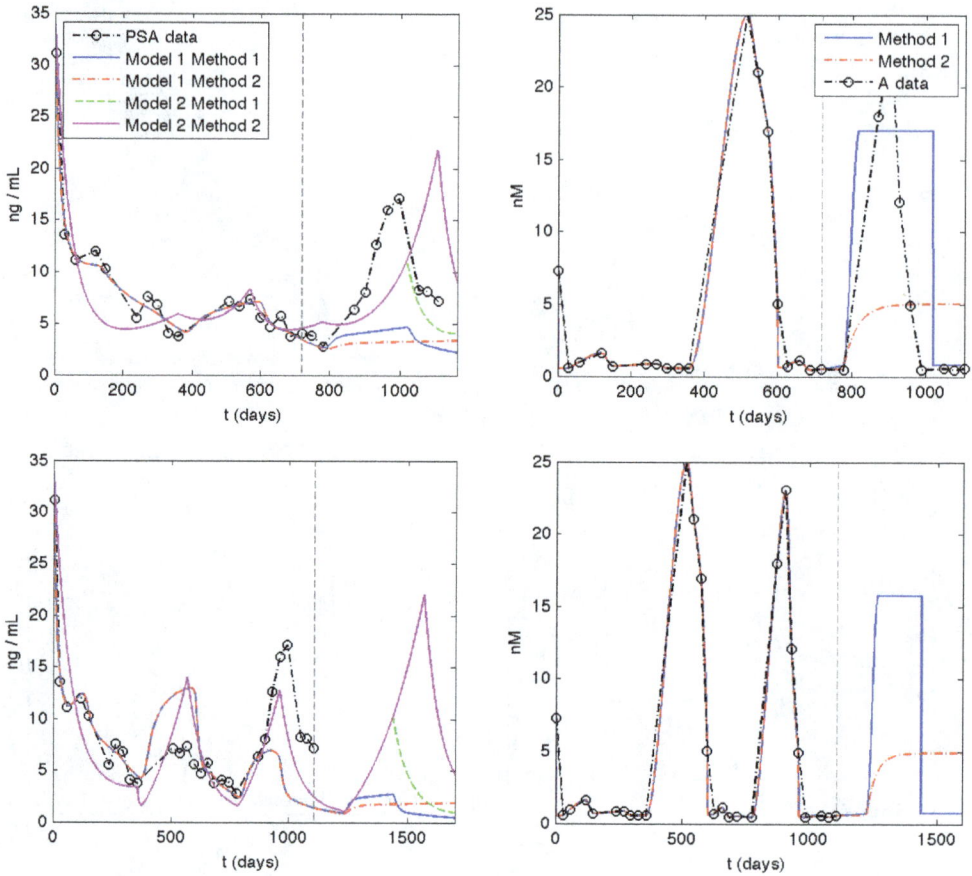

Fig. 4.   Patient 4 PSA levels (left) and serum androgen levels (right) using 1.5 cycles of data (top row) and all 2.5 cycles (second row). The right of the vertical dashed line represents the prediction with the "future" data overlaid for comparison.

repeat this process in order to compare the accuracy of the predictions produced by the two prostate cancer treatment models using the two different predictive methods. The model by Hirata *et al.* is a system of piecewise linear ordinary differential equations representing an AD and two AI cell populations. This model is simpler than the extended Portz model; it contains fewer parameters and does not consider the serum androgen levels. In Method 2, the future serum androgen levels are dependent upon the PSA levels and thus prevent the PSA levels from becoming too high and biologically unreasonable. This method also more accurately follows the methods of the clinical study. While Model 1 using the prediction method proposed by Portz *et al.* (Method 1) is able to fit the used data well, it is not very accurate in predicting the future cycle. When comparing the error values for Method 1 and Method 2 using Model 1, neither method is much more accurate than the other. When comparing Model 1 to Model 2, Model 2 has smaller MSE values for all the

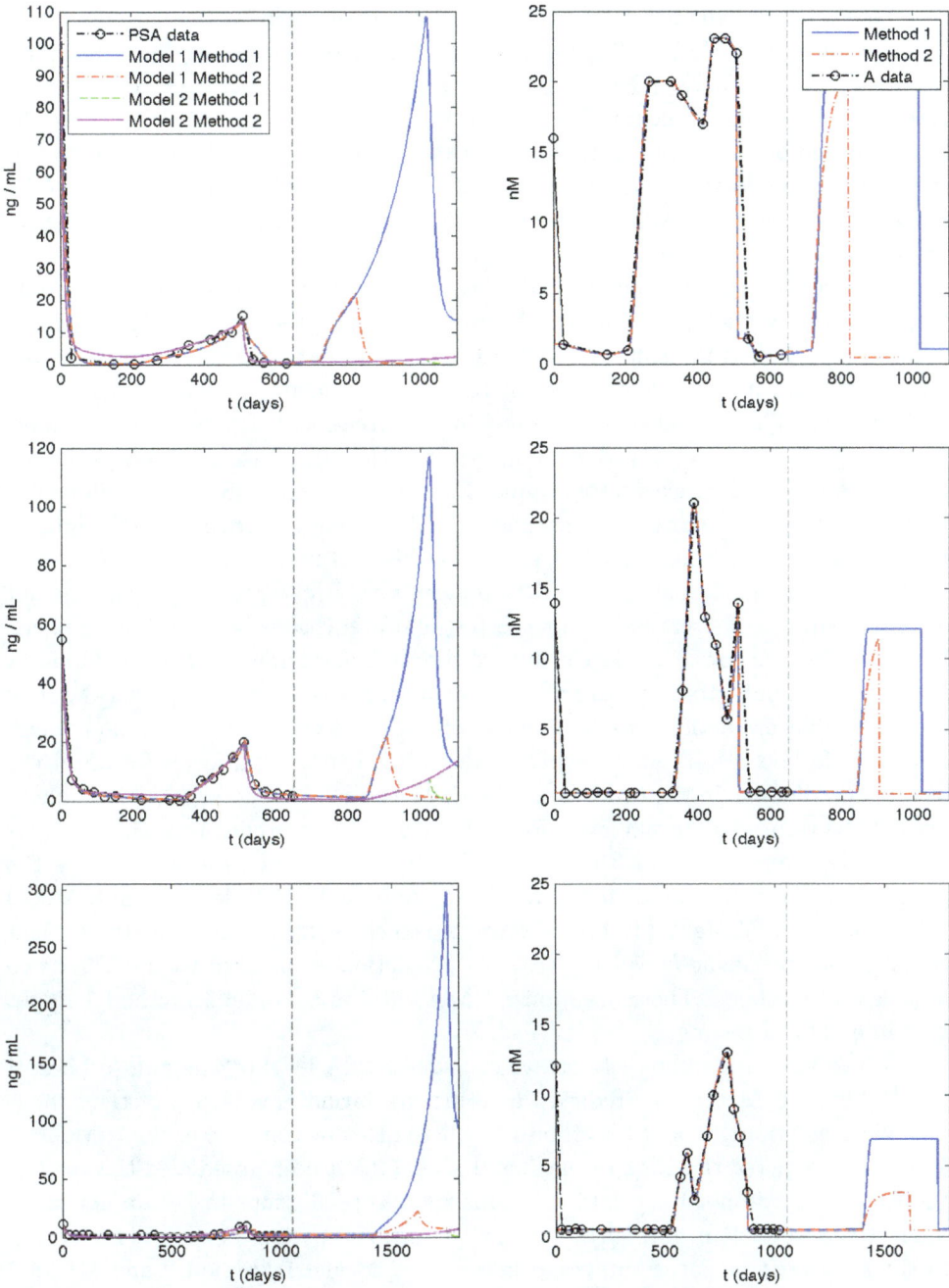

Fig. 5. PSA levels (left) and serum androgen levels (right) for patient 5 (top row), patient 6 (second row), and patient 7 (third row) using all 1.5 cycles of data. The right of the vertical dashed line represents the prediction with the "future" data overlaid for comparison.

testable cycles and smaller MRE values for 2 of the testable cycles. When comparing Method 1 and Method 2 using Model 2, again the methods are comparable, both in MSE and MRE values. Therefore, while neither model is extremely accurate, Model 2 is more accurate in predicting PSA values than Model 1 using the small sample of 7 patients. This implies that while a biologically-based model is important in understanding the biological mechanisms of the process, a simpler model is more accurate and may be more useful for predicting future outcomes of individual patients.

In a clinical setting, a goal of predicting the next cycle of treatment is, not only to accurately predict the future PSA levels, but to determine whether or not a patient can go off-treatment for another cycle, thus improving their health related quality of life. Ideally once the patient resumes treatment, the PSA levels return back to normal levels and remain there while on-treatment. However, it is possible that the patient has developed resistance to treatment and the PSA levels remain higher than normal. Since doctors cannot know if a patient's PSA levels will return back to normal once treatment is resumed, the doctors must use their best judgment to determine if and when a patient should go off-treatment.

We test our mathematical model to determine if it can predict whether or not a patient can go off-treatment. For the testable predictions, we consider "normal levels" to be the low PSA levels during the last on-treatment period. Model 1 Method 1 was only able to correctly predict a return to normal levels for 3 out of the 5 testable predictions in the amount of time shown. Model 1 Method 2 and Model 2 Method 1 were able to predict this return to normal levels for all of the testable predictions in the time shown. With Method 2, the model was only able to predict the return to normal levels for 4 of the 5 testable predictions. However, for 4 of the 12 predicted cycles, the model had not yet predicted the full cycle in the given time; the model predicted a very slow increase in PSA levels. Thus, Model 1 Method 2 and Model 2 Method 1 were able to correctly predict that the patient could go off-treatment for all the testable predictions compared to the other two models and methods. Therefore Model 1 Method 2 and Model 2 Method 1 might help in a clinical setting.

For the nontestable predictions, we consider a PSA level of 4 ng/mL to be normal.[11] Model 1 Method 1 predicted a return to normal levels in 1 of the 7 nontestable predictions, Model 1 Method 2 predicted this return for 5 of the 7, Model 2 Method 1 predicted this for 6 of the 7, and Model 2 Method 2 predicted this for 1 of the 7. Since we do not have data to compare these predictions to, we do not know if the levels return to normal or not, i.e. if the patient has developed resistance to the treatment or not. From the results above, Model 1 Method 2 and Model 2 Method 1 predict that a majority of the patients do not develop resistance in the time shown whereas Model 1 Method 1 and Model 2 Method 2 suggest almost all the patients develop resistance in the time shown.

Mathematical models are important tools for answering biological questions that cannot be answered in a clinical setting. We compare the accuracy of predicted PSA levels using different predicting methods and prostate cancer treatment models. Method 1 followed the predicting method proposed by Portz *et al.*[23] while Method 2 more accurately represented the clinical trial procedure. Since Hirata *et al.*[11,10] also predicted the outcome of patients, we compared an extension of the model by Portz *et al.* (Model 1), a more biologically insightful model, to the model by Hirata *et al.* (Model 2), a simpler model. All the MSE values were smaller for Model 2, but the majority of the MRE values were smaller for Model 1. Both Model 1 Method 2 and Model 2 Method 1 were able to accurately predict that the patient could go off-treatment for every testable prediction, whereas Model 1 Method 1 and Model 2 Method 2 were only able to do so for a subset of the patients. Therefore, while Model 2 may be slightly more accurate, neither model is ready to be used in a clinical setting. This suggests further research of modeling prostate cancer treatment is needed prior to being incorporated into a clinical setting. One possible direction for furthering this research is to reduce the length of predicting time; predicting the next data point instead of next cycle will probably produce more accurate results. As more data is acquired, the parameters can be updated for predicting the following data point. Also, using only the most recent 1.5 cycles of data might produce a more accurate prediction since cancer cells are thought to evolve over time. In order to test this hypothesis, a dataset with several cycles for a patient would be needed. With a larger data set, another direction for future research would be to build a mathematical model which considers the various pathways of resistance and group patients according to different categories, such as stage of cancer, age, and length of treatment.

Acknowledgments

This work is supported in part by NSF DMS-0920744 and the ARCS foundation. We thank Dr. John Nagy and Jason Morken for helpful discussions as well as the anonymous reviewers for their valuable suggestions.

References

1. K. Akakura, N. Bruchovsky, S. L. Goldenberg *et al.*, Effects of intermittent androgen suppression on androgen-dependent tumors, *Cancer* **71**(9), 2782–2790 (1993).
2. NCI Dictionary of Cancer Terms, National Cancer Institute at the National Institutes of Health.
3. R. R. Berges, J. Vukanovic, J. I. Epstein *et al.*, Implication of cell kinetic changes during the progression of human prostatic cancer, *Clinical Cancer Research* **1**(5), 473–480 (1995).
4. J. M. Crook, C. J. O'Callaghan, G. Duncan *et al.*, Intermittent androgen suppression for rising PSA level after radiotherapy, *New England Journal of Medicine* **367**(10), 895–903 (2012).
5. M. R. Droop, 25 years of algal growth kinetics: A personal view, *Botanica Marina* **26**(3), 99–112 (1983).
6. B. J. Feldman and D. Feldman, The development of androgen-independent prostate cancer, *Nature Reviews Cancer* **1**(1), 34–45, October 2001.

7. M. K. Fong, R. Hare and A. Jarkowski, A new era for castrate resistant prostate cancer: A treatment review and update, *Journal of Oncology Pharmacy Practice* **18**(3), 343–354 (2012).

8. M. Gleave, L. Klotz and S. S. Taneja, The continued debate: Intermittent vs. continuous hormonal ablation for metastatic prostate cancer, in *Urologic Oncology: Seminars and Original Investigations*, Vol. 27 (Elsevier, 2009), pp. 81–86.

9. C. S. Higano, Side effects of androgen depreivation therapy: Monitoring and minimizing toxicity, *Urology* **61**(2A), 32–38, February 2003.

10. Y. Hirata, K. Akakura, C. S. Higano, N. Bruchovsky and K. Aihara, Quantitative mathematical modeling of PSA dynamics of prostate cancer patients treated with intermittent androgen suppression, *Journal of Molecular Cell Biology* **4**(3), 127–132 (2012).

11. Y. Hirata, N. Bruchovsky and K. Aihara, Development of a mathematical model that predicts the outcome of hormone therapy for prostate cancer, *Journal of Theoretical Biology* **264**, 517–527 (2010).

12. M. Hussain, C. M. Tangen, D. L. Berry *et al.*, Intermittent versus continuous androgen deprivation in prostate cancer, *New England Journal of Medicine* **368**(14), 1314–1325 (2013).

13. A. M. Ideta, G. Tanaka, T. Takeuchi and K. Aihara, A mathematical model of intermittent androgen suppression for prostate cancer, *Journal of Nonlinear Science* **18** 593–614 (2008).

14. J. T. Isaacs, The biology of hormone refracory prostate cancer: Why does it develop? *Urologic Clinics of North America* **26**(2), 263–273 (1999).

15. T. L. Jackson, A mathematical model of prostate tumor growth and androgen-independent relapse, *Discrete and Continuous Dynamical Systems-Series B* **4**(1), 187–201, February 2004.

16. T. L. Jackson *et al.*, A mathematical investigation of the multiple pathways to recurrent prostate cancer: Comparison with experimental data, *Neoplasia* **6**(6), 697–704 (2004).

17. L. Klotz and P. Toren, Androgen deprivation therapy in advanced prostate cancer: Is intermittent therapy the new standard of care? *Current Oncology* **19**(3), S13–S21, December 2012.

18. F. Labrie, Blockade of testicular and adrenal androgens in prostate cancer treatment, *Nature Reviews Urology* **8**, 73–80, February 2011.

19. J. C. Lagarias, J. A. Reeds, M. H. Wright and P. E. Wright, Convergence properties of the Nelder–Mead simplex method in low dimensions, *SIAM Journal on Optimization* **9**(1), 112–147 (1998).

20. T. Mitin, J. A. Efstathiou and W. U. Shipley, Urological cancer: The benefits of intermittent androgen-deprivation therapy, *Nature Reviews Clinical Oncology* **9**(12), 672–673 (2012).

21. J. D. Morken, A. M. Packer, R. A. Everett, J. D. Nagy and Y. Kuang, Predicting mechanisms of treatment resistance to intermittent androgen deprivation in prostate cancer patients by cell-death rate analysis, submitted, 2013.

22. P. S. Nelson, Molecular states underlying androgen receptor activation: A framework for therapeutics targeting androgen signaling in prostate cancer, *Journal of Clinical Oncology* **30**(6), 644–646 (2012).

23. T. Portz, Y. Kuang and J. D. Nagy, A clincial data validated mathematical model of prostate cancer growth under intermittent androgen suppression therapy, *AIP Advances* **2**, 011002, 1–14 (2012).

24. M. J. Resnick, Urological cancer: Walking the tightrope of survival and quality of life with ADT, *Nature Reviews Clinical Oncology* **10**(6), 307–308 (2013).

25. H. I. Scher, G. Buchanan, W. Gerald, L. M. Butler and W. D. Tilley, Targeting the androgen receptor: Improving outcomes for castration-resistant prostate cancer, *Endocrine-Related Cancer* **11**, 459–476 (2004).

26. M. C. Scholz, R. Y. Lam, S. B. Strum *et al.*, Primary intermittent androgen deprivation as initial therapy for men with newly diagnosed prostate cancer, *Clinical Genitourinary Cancer* **9**(2), 89–94 (2011).

27. N. Sharifi, J. L. Gulley and W. L. Dahut, Androgen deprivation therapy for prostate cancer, *JAMA: The Journal of the American Medical Association* **294**(2), 238–244 (2005).

28. R. Siegel, D. Naishadham and A. Jemal, Cancer statistics, 2013. *CA: A Cancer Journal for Clinicians* **63**(1), 11–30, January/February 2013.

Chapter 10

The Age Specific Incidence Anomaly Suggests that Cancers Originate During Development

James P. Brody

Department of Biomedical Engineering
University of California, Irvine

The accumulation of genetic alterations causes cancers. Since this accumulation takes time, the incidence of most cancers is thought to increase exponentially with age. However, careful measurements of the age-specific incidence show that the specific incidence for many forms of cancer rises with age to a maximum, and then decreases. This decrease in the age-specific incidence with age is an anomaly. Understanding this anomaly should lead to a better understanding of how tumors develop and grow. Here we derive the shape of the age-specific incidence, showing that it should follow the shape of a Weibull distribution. Measurements indicate that the age-specific incidence for colon cancer does indeed follow a Weibull distribution. This analysis leads to the interpretation that for colon cancer two subpopulations exist in the general population: a susceptible population and an immune population. Colon tumors will only occur in the susceptible population. This analysis is consistent with the developmental origins of disease hypothesis and generalizable to many other common forms of cancer.

1. Introduction

Cancers are thought to originate after a series of genetic alterations accumulate in a cell.[1] These alterations could consist of mutations, deletions or modifications to the DNA. The accumulation of alterations increases when certain pathological states occur. A typical colorectal cancer genome contains about a dozen mutated genes that are considered to be driving the cancer.[2,3] Since a normal cell needs to accumulate mutations to more than a dozen key genes before transforming into a tumor cell,[4] the probability of acquiring a particular cancer should increase with age. Thus, it is widely thought that the older one gets, the more likely one is to develop cancer. Age is the primary risk factor for cancer, and cancer is considered an age related disease.[5] Figure 1 shows the textbook understanding of how age relates to cancer.[6]

Figure 1 was initially published in 1980.[7] It was reproduced in a modified form in a very influential review article in 1993.[8] The figure, again slightly modified, appeared in a widely used textbook beginning about 2004.[6] This figure has established in many minds that cancer incidence increases with age.

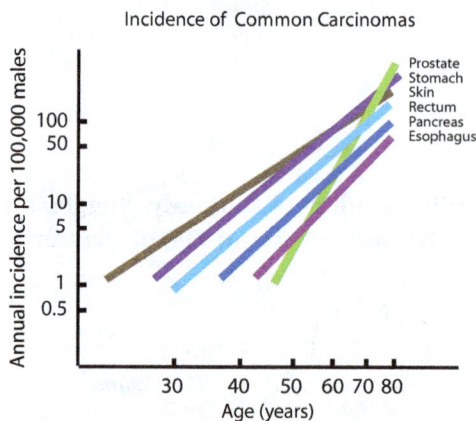

Fig. 1.   (Color online) The textbook illustration of how cancer incidence increases with age. The specific incidence of cancers is often depicted as exponentially increasing with age. An illustration similar to this first appeared in 1980,[7] then in a 1993 review article[8] and eventually in a popular undergraduate molecular biology textbook.[6] This figure has been very influential in suggesting that cancer incidence increases with age.

However, the incidence of most cancers does not monotonically increase with age; instead, the incidence increases to a maximum at some age and then decreases. Although this anomaly is well established,[9–11] it is not widely known.

A complete understanding of the age-specific incidence should lead to a better understanding of how cancers develop. The age-specific incidence is the only quantitative data available on the development process of cancer. It is not confounded by animal models. One of the primary steps to understanding the age-specific incidence is to understand the anomaly.

2. Thought Experiments

We can begin with a *gedanken* experiment. Take 100,000 newborn human infants and put them in an isolated box. These babies live well, and grow into adults. In this idealized experiment, they do not die of any ailments. Each of these experimental subjects is regularly examined for a particular type of cancer, say colon cancer. When a subject is first diagnosed with a colon cancer, the exact age is recorded. The experiment runs for hundreds of years, and then we make a histogram of the number of colon cancers diagnosed as a function of years since the beginning of the experiment.

This histogram will start near zero, reach a maximum then decline to zero again when all subjects have been diagnosed with a first case of colon cancer, as shown in Fig. 2. If every member of the initial population of infants ultimately developed the cancer, then the integral of this histogram should be equal to the population, in this case, 100,000.

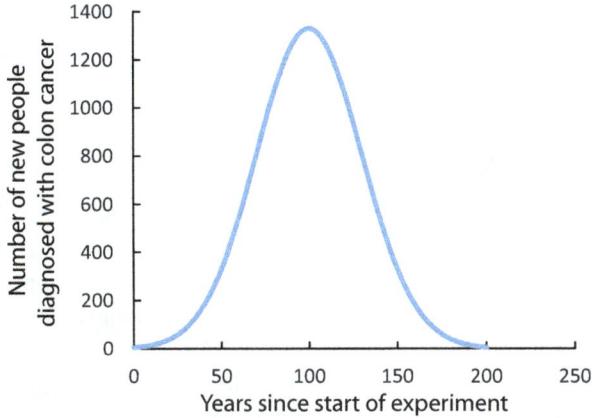

Fig. 2.   The postulated results of a simple thought experiment. The experiment consists of 100,000 newborn humans isolated and monitored annually for colon cancer. The number diagnosed with colon cancer for the first time is recorded each year. After several hundred years, the data is plotted. The plot shows that the incidence increases, reaches a maximum then decreases to zero. The decrease occurs when the majority of the population has already been diagnosed once with colon cancer.

We also perform a second *gedanken* experiment. In this case, the population of 100,000 infants is composed of two apparently indistinguishable subpopulations, one of which can develop colon cancer (20%), and the other of which cannot (80%). The same process is followed. At the end of the experiment, the integral of the histogram is calculated and it will equal 20,000.

From these thought experiments, we can conclude that the number of diagnosed cancers increases with age to some maximum, then decreases. The integral under this curve will be equal to the subpopulation that might develop the particular cancer.

While these *gedanken* experiments are unrealistic, an analogous experiment can be done. First, we record the age of all patients diagnosed with a specific cancer within a large geographic area in one year. Second, we record the age of each member in the entire population. Finally, for each age group, we divide the number of patients who had a tumor diagnosed in that year by the total number of people with that age in the population. By convention, these numbers are multiplied by 100,000 and are called the age-specific incidence.

Strictly speaking, the age-specific incidence is a hazard function. To calculate the hazard function, the population in the quotient above (the total number of people with that age in the population) is reduced by the number of patients with that age who had previously been diagnosed with the specific cancer. Since all cancers only occur in a small fraction of the population, the approximation — that the number of patients with that age who had previously been diagnosed with a specific cancer is much less than the total number of patients with that age — is a very good approximation. Thus, the probability function, as described in the previous paragraph is a good approximation of the hazard function for cancers.

3. Population Based Cancer Registries

Population based cancer registries record the age (and other information) about all patients diagnosed with all types of tumors within a specific geographic area. Then, a government census records the ages of all members of the population within a specific geographic area. Together these sources of data can be combined to compute the age-specific incidence data.

The collection and quality of age-specific incidence data has significantly improved since registries began in the mid 1900's. Initially, this data was derived from death certificates. However, many deaths were attributed to "old age" or non-standard terminology. Today, cancer registries systematically collect information on the diagnosis of tumors and demographic information of the patient.

Different cancer registries, however, collect different information. These differences make aggregation of cancer registry data difficult. In 1973, the National Cancer Institute established the Surveillance, Epidemiology, and End Results (SEER) program. The SEER network of cancer registries solved many of these problems by requiring a specific set of information to be reported and established guidelines on how to encode different properties of a tumor.

The SEER network of cancer registries began with seven different geographic registries covering 16 million people. The program has expanded to 18 cancer registries in 2012, with about 86 million people under surveillance. The SEER program publishes annually case files, which contain summary information about all tumors diagnosed within the specific geographic areas.

Age-specific incidence data collected by the SEER-17 network of cancer registries in 2000 is shown in Fig. 3. This data is presented to emphasize that different cancers have different maximum ages. In each of the seven cancers shown, a decrease in incidence with age exists. This decrease is anomalous, the opposite of the expected behavior, but is consistent with our *gedanken* experiments.

4. Anomaly or Artifact?

A number of concerns have been raised with the surprising observation that the age-specific incidence decreases with age. The three most common are: (1) this observation is contradicted by autopsy studies of latent carcinoma, (2) this observation is an artifact caused by decreased screening rates with age, and (3) this observation is the result of a birth cohort effect. Each of these concerns has been studied in detail and none of these is sufficient to explain the anomaly.

Autopsy studies of latent carcinoma. A widespread perception exists that undiagnosed carcinomas, or latent carcinomas, are common in elderly people. Much

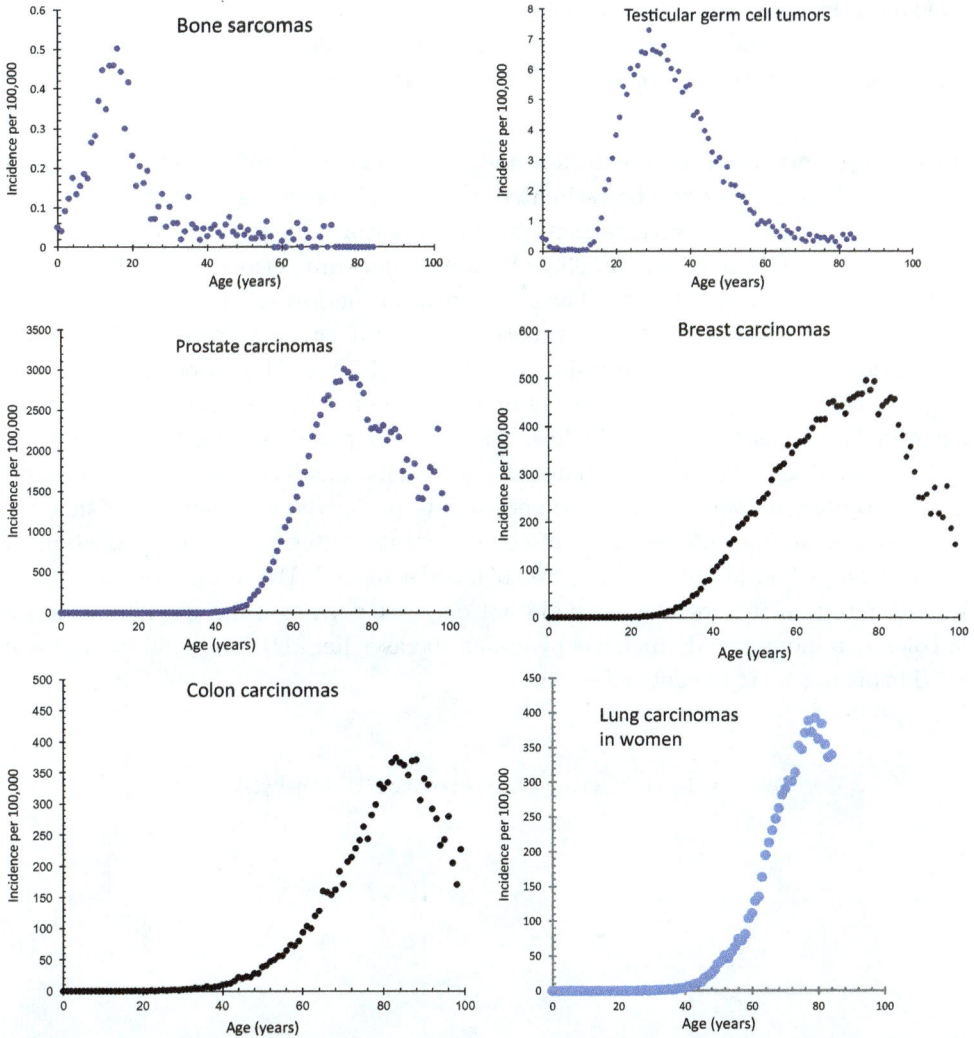

Fig. 3. Many different forms of cancer exhibit the age-specific incidence anomaly. In contrast to the idealized representation in Fig. 1, this figure shows that cancers of the bone, testicles, prostate, breast, colon, and lung all increase with age, reach a maximum, and then decrease. Understanding this anomalous decrease in the age-specific incidence should lead to a better understanding of how cancers begin. These data were collected by the SEER cancer registry[16] and compiled by us. The lung, breast, prostate and colon data were from cases diagnosed between 2000–2010. The bone and testicular cancer data were from cases diagnosed between 1973 and 2011. For the breast, colon, and prostate cancer graphs we estimated the SEER populations for age groups greater than 85 years of age from US Census data as described in Ref. 17.

of this perception is due to work done in the 1950's by LM Franks, who performed several autopsy studies.[12,13] His work appeared to show that many undiagnosed cancers existed in people who died. This work was based on rather small numbers (for instance, Ref. 13 only had two subjects in the 90–100 year old age group).

This data has not been replicated; moreover, recent autopsy studies on larger populations[14,15] contradict Frank's results and have established that the incidence of many common carcinomas *decreases* after a certain age.

Screening and age-specific incidence data. The effect of cancer screening on colorectal cancer rates can be estimated. Colorectal screening rates decrease with age after 60 years. The National Survey of Ambulatory Surgery quantified the rate of outpatient colonoscopies (over 90% of colonoscopies are performed as outpatients) in 1994, 1995, 1996 and 2006.[18] (The survey was not performed in the years between 1996 and 2006.) Based on these estimates, colorectal screening rates in the elderly population (over 85) were about 40% of the rate of 50 to 64 year olds.

The increase in diagnosed cancers due to screening can be estimated from the age-specific incidence data. Guidelines suggest that people should begin screening at 50 years of age. The colon carcinoma age-specific incidence data shows a small, but noticeable, increase over the expected rate at 50 years of age. From this, we estimate the number of new cases of colon carcinoma due to screening at about 2 per 100,000, when about 5000 per 100,000 are screened. Based on these numbers, we estimate that if screening rates did not decrease with age, the specific incidence of colon carcinoma would increase by about 40 cases per 100,000 population at age 85. This is not a significant difference.

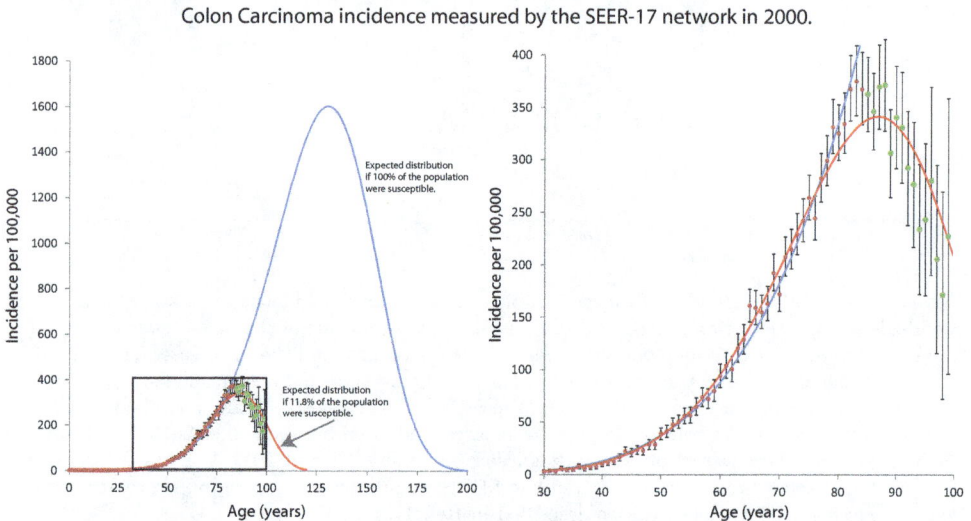

Fig. 4.   (Color online) The age-specific incidence of colon carcinoma, as measured by the SEER-17 network of cancer registries in 2000, is plotted along with the best fit Weibull distribution for 100% of the population and for 11.8% of the population on the left. Both distributions were fit to data from ages 0 to 85 years of age, and then data from 86–99 years of age were plotted. The data is clearly consistent with the 11.8% curve, but not the 100% curve. The plot on the right is a detail of the outlined section of the plot on the left.

Birth cohort effects on age-specific incidence data. The drop in the incidence of colon cancer after age 85 is not due to birth cohort effects. The expected value, if 100% of the population were susceptible, for age 99 is about 850 per 100,000, as shown in data from 2000 in Fig. 4. The observed value is about 227 per 100,000 with a 95% confidence interval of 96 to 357. The observed value is about one quarter of the expected value of colon carcinoma incidence, if 100% were susceptible to colorectal carcinoma. If this population, born in 1900–1910, had a significantly reduced propensity to develop colorectal carcinoma, then we should see a correspondingly small incidence in the population aged 72 years old recorded in 1973 (the earliest SEER data available). No such effect is noticeable in the 1973 data.

Finally, one other objection to the observation of declining cancer incidence with age is sometimes raised mistakenly: competing risk. Competing risk is not relevant here. It is relevant to studies with fixed populations when some members of the population die from other causes. The age-specific incidence data is not based on fixed populations. Specific incidence is the number of diagnosed cancers divided by the population.

5. Understanding the Age-Specific Incidence Data

We follow two approaches to understand the age-specific incidence data. The first approach is theoretical: based on first principles, what should be the shape of the age-specific incidence curve? The second approach asks what biomedical hypothesis could produce age-specific incidence data that we observe.

5.1. *A theory of the age-specific incidence curve*

We have postulated that the age-specific incidence curve should follow an extreme value distribution, in particular the Weibull distribution.[19] Our reasoning is that:

(1) Tumors originate in a single cell, the progenitor cell.
(2) Many potential progenitor tumors cells exist in the body for each type of potential tumor.
(3) A tumor develops when the **first** of these many potential progenitor cells acquires the proper set of mutations.

These steps describe an extreme value process. The Weibull distribution is the proper distribution to characterize this process.[20]

The probability of developing a particular cancer as a function of time, $p(t)$, is given by the Weibull distribution

$$p(t) = A \left(\frac{k}{\lambda}\right) \left(\frac{t - \tau}{\lambda}\right)^{k-1} \exp^{-(\frac{t-\tau}{\lambda})^k}, \qquad (1)$$

when $t \geq \tau$ and the Weibull distribution is $p(t) = 0$ when $t < \tau$. The Weibull distribution has four parameters: A is a normalization factor, τ is the time shift, k

is called the shape parameter, and λ is the scale parameter. Both k and λ must be positive.

We compared the theoretical shape of the age-specific incidence data with observational data collected by the SEER-17 registries in 2000 for colon cancer. We computed the best fit theoretical shape to the 0–84 year old data in two cases. First, when all three parameters (A, k, λ) were allowed to vary and second when A was fixed to be 100,000, but the other two parameters were allowed to vary. These results are shown in Fig. 4.

The two fits in Fig. 4 correspond to the two *gedanken* experiments. If everyone eventually will develop colon cancer, the data points should fall on the 100% curve. We observe the 86–100 years old data points, which were not involved in the fitting, to fall on the 11.8% curve.

This analysis suggests that two subpopulations exist. One subpopulation, consisting of about 12% of the population, is susceptible to developing colon cancer. The second subpopulation, consisting of about 78% of the population is immune to developing colon cancer. Membership in the susceptible population must be determined early, before the age of 20.

One objection to this interpretation is that it apparently contradicts the well-established observation that modifiable risk factors exist for most common cancers. The link between environmental exposure and increased cancer rates is well established, most prominently between cigarette smoking and lung cancer. If environmental exposure causes cancer, how can a susceptible subpopulation exist and be defined at an early age?

One possible explanation is that environmental risk factors affect how fast a tumor grows. For instance, a nonsmoker predisposed to lung cancer might develop a lung tumor at age 120, while a similar heavy smoker develops a tumor at age 60. Since the nonsmoker will probably die from other causes before the lung tumor develops, it appears that the smoker developed lung cancer while the nonsmoker did not develop lung cancer.

The idea that disease observed late in life could originate early in life is not novel. This idea is called the developmental origin of disease hypothesis.[21–23]

5.2. *Developmental origin of disease hypothesis*

Different forms of this hypothesis have been proposed.[24,25] Trichopoulos has suggested that hormonally regulated cancers originate *in utero*. He points out that this would explain a number of curious observations about breast cancer including the dramatic difference in incidence found in Japan and the USA.[26] Barker has suggested that not only cancers, but also other adult diseases have fetal origins.[21,27] Others have also suggested that some chronic diseases are influenced by exposure to environmental factors early in life.[28,29] Diabetes,[30] schizophrenia,[31] and lung disease[32] might also find their origins in early life.

5.3. *Mechanisms for the developmental origins of disease hypothesis*

Several known mechanisms could be responsible for the existence of two subpopulations, these include germ line mutations, somatic mutations early in life, and/or epigenetic modifications inherited or acquired early in life.

Germ line mutations have been ruled out. During the 1990's, significant resources were devoted to the identification of germ line mutations for the most common forms of cancers. This effort led to the identification of BRCA1.[33] Certain mutations in BRCA1 significantly increase the risk that a woman will develop breast cancer. However, these mutations are rare and less than 10% of breast cancers in the US population occur in women with these mutations. Despite searching for similar genes in colon cancer,[34] none has been found with the significance of BRCA1. No recurrent mutations are responsible for the progression of colon cancer.[35]

Somatic mutations acquired early in life (during development) could propagate to encompass entire tissues. Embryonic cells are actively proliferating and a somatic mutation acquired early during development will be found in many cells. Irradiation of a fetus is known to increase the incidence of childhood cancers[36] presumably through the acquisition of somatic mutations. Somatic mutations acquired during development are known to be responsible for retinoblastoma, a childhood cancer.[37]

Epigenetic alterations play a key role in the carcinogenesis process.[38–40] These types of alterations can be passed down through cellular generations. Modification of histones are a key regulatory step in transcription[41] and DNA damage repair.[42] Specific histone modifications have been identified that are common features of human cancers.[43,44] Several approaches to determining genome wide methylation exist, but these approaches have not yet been widely applied to cancer.[45]

6. Conclusion

In conclusion, we showed logically (through *gedanken* experiments), theoretically, and observationally that the age-specific incidence data decreases with age. This apparent anomaly is consistent with the developmental origin of disease hypothesis.

References

1. D. Hanahan and R. A. Weinberg, The hallmarks of cancer, *Cell* **100**, 57–70 (2000).
2. T. Sjöblom *et al.*, The consensus coding sequences of human breast and colorectal cancers, *Science* **314**, 268–274 (2006). http://dx.doi.org/10.1126/science.1133427.
3. L. D. Wood *et al.*, The genomic landscapes of human breast and colorectal cancers, *Science* **318**, 1108–1113 (2007). http://dx.doi.org/10.1126/science.1145720.
4. S. D. Markowitz and M. M. Bertagnolli, Molecular origins of cancer: Molecular basis of colorectal cancer, *N. Engl. J. Med.* **361**, 2449–2460 (2009). http://dx.doi.org/10.1056/NEJMra0804588.
5. J. Campisi and P. Yaswen, Aging and cancer cell biology, *Aging Cell* **8**, 221–225 (2009). http://dx.doi.org/10.1111/j.1474-9726.2009.00475.x.
6. H. Lodish *et al.*, *Molecular Cell Biology*, 5th edn. (W. H. Freeman, 2004).

7. D. G. Miller, On the nature of susceptibility to cancer. The presidential address, *Cancer* **46**, 1307–1318 (1980).
8. B. Vogelstein and K. W. Kinzler, The multistep nature of cancer, *Trends Genet* **9**, 138–141 (1993).
9. C. Harding, F. Pompei and R. Wilson, Peak and decline in cancer incidence, mortality, and prevalence at old ages, *Cancer* **118**, 1371–1386 (2012). http://dx.doi.org/10.1002/cncr.26376.
10. F. Pompei and R. Wilson, Age distribution of cancer: The incidence turnover at old age, *Human and Ecological Risk Assessment* **7**, 1619–1650 (2001).
11. C. Harding, F. Pompei, E. E. Lee and R. Wilson, Cancer suppression at old age, *Cancer Res.* **68**, 4465–4478 (2008). http://dx.doi.org/10.1158/0008-5472.CAN-07-1670.
12. L. M. Franks, Latent carcinoma of the prostate, *J. Pathol. Bacteriol.* **68**, 603–616 (1954).
13. L. M. Franks, Latent carcinoma, *Ann. R. Coll. Surg. Engl.* **15**, 236–249 (1954).
14. K. Imaida *et al.*, Clinicopathological analysis on cancers of autopsy cases in a geriatric hospital, *Pathol. Int.* **47**, 293–300 (1997).
15. J. M. de Rijke *et al.*, Cancer in the very elderly dutch population, *Cancer* **89**, 1121–1133 (2000).
16. Surveillance, Epidemiology, and End Results (SEER) Program Research Data (1973–2010), National Cancer Institute, DCCPS, Surveillance Research Program, Surveillance Systems Branch, released April 2013, based on the November 2012 submission (2013). www.seer.cancer.gov.
17. L. Soto-Ortiz and J. P. Brody, Similarities in the age-specific incidence of colon and testicular cancers, *PLoS One* **8**, e66694 (2013). http://dx.doi.org/10.1371/journal.pone.0066694.
18. K. A. Cullen, M. J. Hall and A. Golosinskiy, Ambulatory surgery in the united states, 2006, *Natl. Health Stat. Report*, 1–25 (2009).
19. L. Soto-Ortiz and J. Brody, A theory of the cancer age-specific incidence data based on extreme value distributions, *AIP Advances* **2**, 011205 (2012).
20. W. Weibull, A statistical distribution function of wide applicability, *ASME Journal of Applied Mechanics* **18**, 293–297 (1951).
21. D. J. Barker, The fetal and infant origins of adult disease, *BMJ* **301**, 1111 (1990).
22. D. J. P. Barker, The developmental origins of adult disease, *Eur. J. Epidemiol.* **18**, 733–736 (2003).
23. D. J. P. Barker, J. G. Eriksson, T. Forsén and C. Osmond, Fetal origins of adult disease: Strength of effects and biological basis, *Int. J. Epidemiol.* **31**, 1235–1239 (2002).
24. S. Morgenthaler, P. Herrero and W. G. Thilly, Multistage carcinogenesis and the fraction at risk, *J. Math. Biol.* **49**, 455–467 (2004). http://dx.doi.org/10.1007/s00285-004-0271-9.
25. P. Herrero-Jimenez *et al.*, Mutation, cell kinetics, and subpopulations at risk for colon cancer in the United States, *Mutat. Res.* **400**, 553–578 (1998).
26. D. Trichopoulos, Hypothesis: Does breast cancer originate in utero? *Lancet* **335**, 939–940 (1990).
27. K. Calkins and S. U. Devaskar, Fetal origins of adult disease, *Curr. Probl. Pediatr. Adolesc. Health Care* **41**, 158–176 (2011). http://dx.doi.org/10.1016/j.cppeds.2011.01.001.
28. P. D. Gluckman and M. A. Hanson, Living with the past: Evolution, development, and patterns of disease, *Science* **305**, 1733–1736 (2004). http://dx.doi.org/10.1126/science.1095292.

29. P. D. Gluckman, M. A. Hanson, C. Cooper and K. L. Thornburg, Effect of *in utero* and early-life conditions on adult health and disease, *N. Engl. J. Med.* **359**, 61–73 (2008). http://dx.doi.org/10.1056/NEJMra0708473.

30. C. S. Yajnik, Early life origins of insulin resistance and type 2 diabetes in india and other Asian countries, *J. Nutr.* **134**, 205–210 (2004).

31. D. St Clair *et al.*, Rates of adult schizophrenia following prenatal exposure to the chinese famine of 1959–1961. *JAMA* **294**, 557–562 (2005). http://dx.doi.org/10.1001/jama.294.5.557.

32. R. Harding and G. Maritz, Maternal and fetal origins of lung disease in adulthood, *Semin. Fetal Neonatal Med.* **17**, 67–72 (2012). http://dx.doi.org/10.1016/j.siny.2012.01.005.

33. Y. Miki *et al.*, A strong candidate for the breast and ovarian cancer susceptibility gene BRCA1, *Science* **266**, 66–71 (1994).

34. P. Peltomaki *et al.*, Genetic mapping of a locus predisposing to human colorectal cancer, *Science* **260**, 810–812 (1993).

35. A. P. Feinberg, R. Ohlsson and S. Henikoff, The epigenetic progenitor origin of human cancer, *Nat. Rev. Genet.* **7**, 21–33 (2006). http://dx.doi.org/10.1038/nrg1748.

36. R. Doll and R. Wakeford, Risk of childhood cancer from fetal irradiation, *Br. J. Radiol.* **70**, 130–139 (1997).

37. S. A. Frank and M. A. Nowak, Cell biology: Developmental predisposition to cancer, *Nature* **422**, 494 (2003). http://dx.doi.org/10.1038/422494a.

38. M. Esteller, Epigenetics in cancer, *New England Journal of Medicine* **358**, 1148–1159 (2008).

39. R. L. Jirtle, Genomic imprinting and cancer, *Exp. Cell. Res.* **248**, 18–24 (1999). http://dx.doi.org/10.1006/excr.1999.4453.

40. R. L. Jirtle and M. K. Skinner, Environmental epigenomics and disease susceptibility, *Nat. Rev. Genet.* **8**, 253–262 (2007). http://dx.doi.org/10.1038/nrg2045.

41. P. A. Jones and S. B. Baylin, The epigenomics of cancer, *Cell* **128**, 683–692 (2007). http://dx.doi.org/10.1016/j.cell.2007.01.029.

42. P. Chi, C. D. Allis and G. G. Wang, Covalent histone modifications — Miswritten, misinterpreted and mis-erased in human cancers, *Nat. Rev. Cancer* **10**, 457–469 (2010). http://dx.doi.org/10.1038/nrc2876.

43. M. F. Fraga *et al.*, Loss of acetylation at lys16 and trimethylation at lys20 of histone h4 is a common hallmark of human cancer, *Nat. Genet.* **37**, 391–400 (2005). http://dx.doi.org/10.1038/ng1531.

44. C. Das, M. S. Lucia, K. C. Hansen and J. K. Tyler, Cbp/p300-mediated acetylation of histone h3 on lysine 56, *Nature* **459**, 113–117 (2009). http://dx.doi.org/10.1038/nature07861.

45. P. Laird, Principles and challenges of genome-wide DNA methylation analysis, *Nature Reviews Genetics* **11**, 191–203 (2010).

Chapter 11

Cancer — Pathological Breakdown of Coherent Energy States

Jiří Pokorný

Institute of Photonics and Electronics
Academy of Sciences of the Czech Republic
Chaberská 57, Prague 8–Kobylisy
182 51, Czech Republic
pokorny@ufe.cz

Jan Pokorný

Institute of Physics
Academy of Sciences of the Czech Republic
Na Slovance 2, Prague 8
182 21, Czech Republic
pokorny@fzu.cz

Jitka Kobilková

1st Faculty of Medicine
Department of Obstetrics and Gynaecology
Charles University, Apolinářská 18, Prague 2
128 00, Czech Republic
jitka.kobilkova@centrum.cz

Anna Jandová

Senior Private Scientist
(previously Institute of Photonics and Electronics
Academy of Sciences of the Czech Republic)
Tyršova 415, Šestajovice, 250 92, Czech Republic

Jan Vrba

Faculty of Electrical Engineering
Czech Technical University in Prague
Technická 2, Prague 6, 166 27, Czech Republic
vrba@fel.cvut.cz

Jan Vrba Jr.

Faculty of Biomedical Engineering
Czech Technical University in Kladno
Sitná Square 3105, Kladno
272 01, Czech Republic
jan.vrba@fbmi.cvut.cz

The fundamental property of biological systems is a coherent state far from thermodynamic equilibrium excited and sustained by energy supply. Mitochondria in eukaryotic cells produce energy and form conditions for excitation of oscillations in microtubules. Microtubule polar oscillations generate a coherent state far from thermodynamic equilibrium which makes possible cooperation of cells in the tissue. Mitochondrial dysfunction (the Warburg effect) in cancer development breaks down energy of the coherent state far from thermodynamic equilibrium and excludes the afflicted cell from the ordered multicellular tissue system. Cancer lowering of energy and coherence of the state far from thermodynamic equilibrium is the biggest difference from the healthy cells. Cancer treatment should target mitochondrial dysfunction to restore the coherent state far from thermodynamic equilibrium, apoptotic pathway, and subordination of the cell in the tissue. A vast variety of genetic changes and other disturbances in different cancers can result in several triggers of mitochondrial dysfunction. In cancers with the Warburg effect, mitochondrial dysfunction can be treated by inhibition of four isoforms of pyruvate dehydrogenase kinases. Treatment of the reverse Warburg effect cancers would be more complicated. Disturbances of cellular electromagnetic activity by conducting and asbestos fibers present a special problem of treatment.

1. Introduction

Biological systems are complex structures organized from elementary mass units containing individual parts at different hierarchical levels. Properties of the parts and of the whole system are created by the structural organization. A mammalian body is arranged from elementary living units — cells created during embryo development after fertilization of the egg. The fertilized egg cleaves and forms many small cells, then a basic body plan is created, the rudiments of organs set up, and tiny organs to the adult shape formed. All cells and their subunits are built from atoms and molecules which are not living forms themselves. The living state is established in the composed system. After inclusion of new macromolecules and particles into the cell they become a part of the living system. Transformation of the composed structures into living state is an essential question. The mature egg in a mammalian ovary is a living system. Physical processes could establish life in structures created on the basis of chemical reactions and chemical binding. It is known that energy is continuously supplied to any biological system. Any component of the biological activity (for instance transport, organization, motion, brain activity etc.) depends on energy supply. Energy transformation processes and energy excitations are inseparable parts of living systems. The energy supply creates and sustains a state far from the thermodynamic equilibrium which is considered to be the basis of life. Formation of this state is conditioned by low energy losses by damping, emission, and parasitic consumption. The state far from the thermodynamic equilibrium very likely impresses a pattern of non-random coherent activity correlated in space and time in the biological system. As the pattern of correlation and coherence is expressed in the whole biological system regardless of its dimensions, a long-range mechanism based on physical forces is assumed. Due to an exceptional electric polarity of components and structures of living cells, the acting forces are assumed to be of electrodynamic and electromagnetic origin. The generation processes seem to depend on frequency region. Nonlinear interaction between elastic and polarization fields with random excitations could generate electrodynamic activity in low frequency bands (these processes could be combined with free charge oscillations). Photons released from chemical reaction are important for excitation in UV and visible range. Electrodynamic activity is a fundamental property of biological systems performing energy transformation for mechanical work and information transfer.

Biological systems are dependent on the ambient medium. They interact with the surroundings, uptake mass, energy, and obtain information from it. Each biological system detects its difference from the surroundings based at least on sensing random or unfamiliar coherent signals. Biological systems evaluate their difference from the medium around them. There are further properties of living entities. Even very simple systems like viruses strive for continuation in time of their own entity and/or their descendants. Biological systems endeavor to provide the most convenient conditions for their existence to avoid disturbances endangering their normal state or their existence. Pathological states may be caused by disturbances of any part and/or activity of the complex system, in particular of material, organization, and the state far from the thermodynamic equilibrium. Diseases based on pathological defects of cellular energy systems were experimentally studied by Jandová *et al.*[1]

It is well known that the majority of proteins and protein structures are electrically polar. They are electric dipoles or multipoles and any vibration generates electromagnetic field. Consequently, the biological activity should depend not only on the biochemical–genetic processes but also on the biophysical mechanisms with the dominant role of the electromagnetic field. Research on the electromagnetic activity of biological systems was initiated by H. Fröhlich at the first Versailles conference on Theoretical Physics and Biology in 1967.[2] Fröhlich formulated a hypothesis of a strong excitation of one or a few modes of motion, stabilized due to low emission and friction losses, phase correlated over macroscopic regions, and superimposed on random thermal fluctuations. The strong electric polarity of biological objects suggested longitudinal electric oscillations as stabilizing modes. Taking into account the physical principles of the electrodynamic activity, Fröhlich[3–6] formulated possible mechanisms based on nonlinear interactions between longitudinal elastic and electric polarization fields, energy transfer between normal modes along frequency scale, and energy condensation in the lowest frequency mode.

Fröhlich's hypothesis laid the basis for understanding physical processes in biological systems. Fröhlich also assumed that the cancer transformation pathway includes a link with altered coherent electric vibrations. A cancer cell may escape from interactions with the surrounding healthy cells and may perform individual independent activity if the frequency spectrum is rebuilt and shifted.[7] The frequency changes may be combined with disturbances of the spatial pattern of the field. The transformed cell is released from local interactions and prepared to undergo local invasion and formation of metastases. Fröhlich was ahead of his time. Biological research at that period was orientated towards the chemical reaction and the genetic code problems and the Fröhlich's hypothesis was not considered to have any biological significance.

Fröhlich's hypothesis of polar modes, generation of the electromagnetic field, and its role in biological activity and the cancer transformation is a logical continuation of Warburg's discovery of partial suppression of the oxidative metabolism in cancer tissues. O. Warburg intuitively assessed the cancer transformation as a disturbance of the energy processing system. He experimentally proved that cells from a cancer tissue can obtain approximately the same amount of energy from fermentation as from oxidation, whereas healthy cells obtain much more energy from oxidation than from fermentation.[8,9] In the time of Warburg's life the defect of oxidative metabolism was considered to be a side effect, rather than the main point of the cancer process. Most malignant tumors have an increased glucose uptake as was disclosed by positron emission tomography (PET) imaging and published by Bonnet *et al.*,[10] which is consistent with the metabolic phenotype of the aerobic glycolysis described by Warburg. The decreased oxidative metabolism is caused by dysfunction of mitochondria in cancer cells. Recently, a modified version of the Warburg effect was revealed by Pavlides *et al.*[11] The cancer cell (for instance, an epithelial breast cancer cell) has a fully functional mitochondria and the mitochondrial dysfunction is transferred to fibroblasts associated with the cancer cell. Fibroblasts supply energy rich metabolites to the cancer cell. This type of cancer is connected with the term the reverse Warburg effect.

Mitochondria form a boundary between biochemical–genetic and physical processes of energy transformation and utilization. Chemical signals switch the function of pyruvate dehydrogenase complex on and off and in this way regulate the mitochondrial function and the physical processes dependent on energy supply. The role of mitochondrial dysfunction in cancer development is central for the creation of malignancy. In a cervical cancer the mitochondrial dysfunction is formed in the transformation link from precancerous to cancer cells as was experimentally studied by Jandová *et al.*[12] Therefore, in cervical cancers local invasion and the process of metastases develop after establishment of the mitochondrial dysfunction. Degradation of the mitochondrial function is at the beginning of cancer generalization.

This paper analyses and describes the role of mitochondria in generation of the polar oscillations in microtubules, their role in biological activity, excitation and maintenance of the coherent state far from the thermodynamic equilibrium. Studies of cancer disturbances of the coherent state far from the thermodynamic equilibrium as a central cancer problem are included.

2. Electromagnetic Activity of Living Cells

Electric and electromagnetic oscillations have been measured on living cells. Pohl, Pohl *et al.*, and Roy *et al.*[13–15] observed attraction of small dielectric particles to living cells and assessed the corresponding frequency of oscillations in the range below 10 MHz. Pohl explained the observed phenomenon by dielectrophoresis.[16] The greater the permittivity of the particles and the smaller the conductivity of the cellular suspension, the greater the number of attracted particles. Beside dielectrophoretic measurements of yeast and alga cells, Hölzel and Lamprecht and Hölzel[17,18] also measured oscillations in the frequency range 1.5–52 MHz using a special detection and amplification system. Electric oscillations of yeast cells in the range 8–9 MHz were measured by Pokorný *et al.*[19] Mechanical vibrations of the membranes of yeast cells in the acoustic frequency region were measured by Pelling *et al.*[20,21] by AFM (Atomic Force Microscope). The mechanical vibrations measured by AFM were compared with the electric oscillations detected at the yeast cell membranes.[22] Damping of the external electromagnetic field by the cancer tissue at the frequency 465 MHz and its first harmonic was measured by Vedruccio and Meessen.[23] Electromagnetic field generated by living cells in the red and near-infrared region and causing interaction between them was measured by Albrecht-Buehler.[24−26] Cells also detect electromagnetic signals and send pseudopodia to the source. The photon emission from living bodies was measured for instance by Popp.[27] The experimental results suggest that eucaryotic living cells can generate electromagnetic field in a wide frequency region.

3. Oscillations in Microtubules

Microtubules form a part of a well-organized cytoskeleton structure. Cytoskeleton is a specific filamentous network in eucaryotic cells exerting forces and generating movements without any major chemical change. Eucaryotic cells can form multicellular structures and systems. Eucaryotic cell's capability to create multicellular organisms depends on communication and cohesion between the cells.

The structures generating the electromagnetic field have to be electrically polar, nonlinear, and excited by energy supply. Analyses of properties of such cellular structures were published after Fröhlich's death. Fröhlich[6] assumed generation by the plasma membrane. Microtubule filaments — hollow tubes with the inner and outer diameter 17 and 25 nm, respectively — were described by Amos and Klug.[28] Microtubules grow from the centrosome in the center of the cell and form a radial cellular structure which is the main organizer of the cytoskeleton. Large dielectrophoretic effects of the yeast cells in the M phase (when cells divide and the microtubule activity is high) were measured by Pohl *et al.*[14] Tuszyński *et al.*[29] proved that heterodimers in microtubules are electric dipoles. Generation of the electromagnetic field by microtubules based on Fröhlich's mechanism of the electrical polar vibrations was proposed by Pokorný *et al.*[30] Electric oscillations measured at the cellular membrane of living yeast cells in the M phase display enhanced electric activity in some periods coinciding with mitotic spindle formation, metaphase, and anaphase A and B.[19] Disruption of microtubule polymerization in cells by an external electromagnetic field at the frequency 0.1–0.3 MHz suggests microtubule electromagnetic activity in heterodimer attachment.[31]

Resonant frequencies of microtubules were measured in the frequency range of 10–30 MHz and 100–200 MHz by Sahu *et al.*[32] The resonant frequencies were disclosed by measurement of DC conductivity after application of the oscillating electromagnetic signal and from transmittance and reflectance of microtubules without and with a compensation of parasitic reactances in the frequency range of 1 kHz–1.3 GHz. The resonant frequencies do not depend on the length of the microtubule. After release of water from the microtubule cavity, the peaks of resonance are not observed.

The experimental results proved that microtubules form resonant oscillating circuits. Nonlinear properties of microtubules make possible transformation of the energy of oscillations between different frequency regions. If the energy supply is sufficiently high a coherent state may be formed. The water core inside the microtubule resonantly integrates all the heterodimers in such a way that the microtubule nanotube functions like a single heterodimer irrespective of the microtubule size. The enhanced electrodynamic activity of the cells in the M phase corresponds to the development and functions of the mitotic spindle. Therefore, the experimental data support the idea that microtubules are generators of the electromagnetic activity in living cells. However, direct measurement of a single microtubule in a living cell has not been performed yet. The physical mechanism of microtubule oscillations and generation of the field are not fully explained. The polar modes may interact with the free charges and the water molecules inside the microtubule. The resonant frequencies may also depend on electron oscillations in the secondary structure of heterodimers. Nevertheless, interaction between the elastic and electric oscillations seems to be important.

Several mechanisms are utilized for the energy supply to microtubules. The energy is supplied by hydrolysis of GTP to GDP in β tubulins after polymerization,[30,33] motion of motor proteins along microtubules,[20] and very likely also by nonutilized energy liberated from mitochondria.[34,35] Chemical reactions release photons and in this way may supply the energy to oscillations in the UV and visible wavelength regions.

Microtubule oscillations below 1 GHz very likely form only a low frequency component of the whole biological electromagnetic spectral range. Some parts of biological systems may represent resonant circuits which may be excited. A living cell forms a cavity resonator for electromagnetic waves at the frequency of about 10^{13} Hz, which corresponds to a cell of a spherical shape with a diameter about 10 μm.[36] The positions of the mitotic spindle poles may correspond to the nodes of the cavity electromagnetic field and the geometrical shape of polar and kinetochore microtubules to the line of force of the field.[27] Dimensions of the inner microtubule cavity correspond to a soft X-ray resonator.[36]

4. Mitochondria Support Microtubule Oscillations

Mitochondria are multifunctional organelles in the cell. They have different shapes with linear dimension of about 0.5–1 μm and occupy a substantial portion of the cytoplasmic volume of eucaryotic cells. The activity of mitochondria is provided on their inner membrane. The energy released from the foodstuffs is parceled out by mitochondria with utilization of oxygen into small packets for efficient covering of biological needs (the oxidative metabolism of mitochondria). However, mitochondrial function encompasses several fundamental physical processes and cannot be reduced to mere production of ATP and GTP. Utilization of the chemical energy for proton transfer from the mitochondrial matrix to the intermembrane space and proton diffusion into cytosol through holes in the outer membrane is an important intermediate mechanism in the energy production. A layer of a strong static electric field created around mitochondria up to a distance of several micrometers was measured by Tyner *et al.*[37] The static electric field changes the phase of water. Layers of ordered water are formed. Water ordering by a strong electric field is a general phenomenon in nature; water is ordered around charged surfaces. Special layers around microtubules 5–20 nm thick (clear zones) were measured by Amos.[38] Formation of the clear zones was assumed to depend on the negative electrostatic charge at the microtubule surface.[39] The interfacial ordering was studied and described by Zheng *et al.*,[40] Pollack *et al.*,[41] Chai *et al.*,[42,43] and Pollack.[44] Fuchs *et al.*[45–47] and Giuliani *et al.*[48] investigated formation of a floating water bridge between two glass beakers after application of the field by electrodes. The ordered water exhibits separation of charges and loses its viscous damping property.[49,50] Explanation of these findings based on the theory of arrangement of coherent microscopic domains (that exist in water) into macroscopic ordered layers was published by Preparata[51] and Del Giudice and Tadeschi.[52]

A periodic character of the cell development is a general process in nature. The cell cycle has distinct phases. The M phase denotes the process of nuclear division and separation into two cells. The remaining portions of the cell cycle are included into interphase, i.e. the period between two M phases. In the interphase mitochondria are aligned along microtubules in the regions of greatest energy consumption. The strong static electric field around mitochondria may shift oscillations in molecules and structures into a highly nonlinear region.[49,50] Significant reduction of the water viscosity damping of microtubule oscillations is caused by the ordered water around them.[49] Analysis of the effect of a protective layer of the ordered water on damping was performed by Pokorný.[53]

5. Interactions Between Cells

The electromagnetic fields generated by living cells in the frequency region below about 100 MHz may be important in organization of tissues, synchronization of biological processes, and establishment of coherence in low frequency bands. The dimensions of the biological bodies and particular organs are much smaller than the wavelength of the field. In a medium with relative permittivity about 100, wavelengths at the frequencies 10 kHz and 10 MHz are 3000 m and 3 m, respectively. However, the electromagnetic field generated by a cell depends on the spatial arrangement of microtubules and the way of their excitation in the cell. In an ideal case microtubules form a discrete spherically symmetrical structure. If the oscillations correspond also to spherical symmetry, only a weak electromagnetic field is generated in the direction of microtubule axes at a short distance from the plasma membrane. The polar modes generated around cells by the plasma membrane can mediate interactions between the cells which are not in a direct contact. The discrete spherical symmetry of microtubule oscillations could be disturbed if the cells are in direct contact. The excited microtubule oscillations can generate a tissue field. Due to character of the near field the high intensity of the longitudinal components of the electric field of the microtubule dipoles might be dominant in interactions. The interaction energy was analyzed to assess interactions between oscillating systems. The interaction forces are very small if the frequencies of interacting cells are different as was evaluated by Fröhlich[7] and Pokorný and Wu.[54] The effect of coupling of the Fröhlich's polar modes to the heat bath on the interaction forces was analyzed by Pokorný.[35,55] The interactions are spectral sensitive.

The motor proteins provide essential transport along the microtubules. The directional transport of biological molecules and reaction components in the cytosol cannot be explained on the basis of a random Brownian motion. A combination of electrodynamic deterministic and random forces in the directional transport of mass particles was analyzed.[56] Organization of the living matter might depend on the directional transport.[57] Communication between the brain and various parts of the body could be mediated by streams of photons, which may provide a high capacity information transfer.[58]

6. Cancer Process

Cancer is a multistep and multibranch microevolutionary process. The links that compose the cancer transformation pathway contain disturbances of the biochemical–genetic and physical origin. At the cancer beginning the links comprise a wide spectrum of processes of different nature. Besides chemical and genetic changes caused by different agents, also mechanotransduction[59,60] disturbances may develop into cancer. Therefore, the classification and treatment of cancers based on the processes in the initial links is a complex task. Above it, the cells in their development can change their genetic make-up. But all these processes at a critical stage of development trigger the defect of the oxidative metabolism caused by inhibition of the pyruvate transfer into the mitochondrial matrix.[10] Warburg's experimental research disclosed that all measured cancer tissues displayed the mitochondrial dysfunction.[8] The mitochondrial dysfunction was observed in cancer cells or in the fibroblasts associated with a cancer cell.[11] The experimental results of Jandová et al.[12] suggest that the mitochondrial dysfunction in cervical carcinoma develops in precancerous state. The mitochondrial dysfunction results in the changed energy production and altered behavior of cells, in particular disturbed interaction with other cells leading to local invasion and beginning of tumor generalization. This state is assigned to disturbances of the energy coherent states. The origin of cancer is a problem which is not yet solved. It is assumed that several signal triggers producing cancer have to act on the cell. Changed genes are observed in the initial links before malignant properties appear. Ionic radiation, external mechanical forces, or other external agents may result in a dangerous cancer triggering signal. A mechanical transduction of external pressure or tension can cause the genetic changes and mitochondrial dysfunction too.[59,60] It should be mentioned that cancer might be also set up by chronic decrease of electromagnetic activity causing inaccurate or wrong cellular mechanisms. A unique cancer triggering mechanism has not been revealed yet.

The mitochondrial dysfunction has two modifications which determine the normal and the reverse Warburg effects; transfer of pyruvate to the mitochondrial matrix is inhibited in the cancer cells or in fibroblasts associated with the cancer cells, respectively. There are a few biochemical molecules blocking the pyruvate transfer. In the cancer cells with the normal Warburg effect the pyruvate dehydrogenase complex in mitochondria is regulated by pyruvate dehydrogenase kinases (PDK). This type of cancer cells termed the glycolytic phenotype was described by Bonnet *et al.*[10] McFate *et al.*[61] published that there are four isoforms PDK-1–4. Inhibition of pyruvate processing by mitochondria causes disturbances of the static electric field and the ordered water layer around them leading to increased damping of the microtubule oscillations. The power of microtubule oscillations is lowered and their frequency altered. The interaction forces of the cancer cells with the healthy cells in the tissue are reduced and conditions for generalization of the tumor are prepared.

The Warburg effect link along the cancer transformation pathway is assumed to precede the malignant properties. The idea is based on measurement of the response of the cell mediated immunity to the antigen of lactate dehydrogenase elevating virus (LDV). LDV enhances the level of LDH isoenzymes. The antigen was prepared from serum of inbred mice C3H H^{2k} strain infected with the LDH virus. The response to the antigen was investigated by Jandová *et al.*[12] T lymphocytes were prepared from venous blood of healthy women, patients with cancer and precancerous lesions of cervix. Effects of the LDH virus and the cervical cancer antigen are similar. The results suggest that the mitochondrial dysfunction in the cervical cancer development is caused in the precancerous link of cancer transformation. The mitochondrial dysfunction seems to be an essential condition for the malignant activity of cancer cells.

The cancers with the normal and the reverse Warburg effect form two main groups of cancers. The mitochondrial dysfunction is set up in cancer cells or in their associated fibroblasts. Cancer cells with mitochondrial dysfunction produce only about one half of the cell energy production by the oxidative metabolism. In healthy cells the oxidative ATP production may be even 100 times greater than the fermentative one. Damadian[62] found by measurement of the nuclear magnetic resonance (NMR) that cancer cells create a less ordered system. He wrote that "the malignant tissues were characterized by an increase in the motional freedom of tissue water molecules". In contrast, Kiricuta and Simplăceanu claimed that the main cause of the differences observed between the spin–lattice and spin–spin relaxation times of the normal and malignant tissue is the higher water content in the latter tissue.[63] NMR relaxation times are characteristic properties that depend on the physical processes of water in the measured system and about 20% increase in volume of water cannot change them. Assuming that the cells contain both the ordered and the bulk water, then the relaxation times are described by the sums of weighted exponential components representing both phases. The differences in the relaxation times between healthy and cancer tissues are about 300%. The measured differences in water content cannot cause significant changes. The main difference results from the viscosity of the ordered and the bulk water. The viscosity of the bulk water with low level of ordering causes large values of the relaxation times and damping of oscillations in microtubules.[49] The diminished static electric field around mitochondria might result in a shift of the microtubule oscillations toward a linear region. Consequently, power of the electrodynamic field is lowered, the coherence diminished, and the frequency spectrum shifted and rebuilt.

Cancers with the reverse Warburg effect were observed by Pavlides et al.[11] at breast cancer cells. Since then properties of the novel phenotype of cancers have been described in a large amount of publications (some references are included[64–74]). The epithelial breast cancer cells have fully functional mitochondria and the mitochondrial dysfunction is induced in associated stromal fibroblasts. This pathological process is conditioned by a loss of expression of caveolin-1 in the stroma. The energy rich metabolites (pyruvate, lactate, glutamine, ADMA-asymmetric dimethyl arginine, and BHB — beta-hydroxybutyrate) produced by fermentation are supplied to a cancer cell from the associated fibroblasts with dysfunctional mitochondria. The energy production and power of the electrodynamic field in the cancer cells are high. The microtubule oscillations are shifted to a highly nonlinear region and the frequency spectrum is changed. The energy rich supply from the environment support growth of the cancer cell and its aggressiveness.

The two types of cancers were known and distinguished about thirty years ago by measurement of potential difference of the mitochondrial inner membrane (but these different types of cancers were not connected with the Warburg effect). The membrane potential (negative inside) depends on the distribution of the negative and positive charges connected with the mitochondrial function. Positively charged protons are transferred across the inner membrane. The measurements were performed by a fluorescent method — uptake and retention of positively charged fluorescent dyes (URFD). For instance, the high value of URFD (called hyperpolarization) was measured at a large amount of tumors of ovary, kidney, colon, liver, and other organs showing that a great majority of carcinomas and melanomas are of the glycolytic phenotype.[75,76] The difference in the mitochondrial membrane potential between the normal cells and the carcinoma cells of at least 60 mV was measured by Modica-Napolitano and Aprille.[77] On the other hand the most significant exceptions have been oat and large cell carcinomas of lung, poorly differentiated carcinoma of colon, lymphomas, sarcomas, and neuroblastomas,[75,76] where low values of URFD were measured. Bonnet *et al.*[10] proved that after treatment of the human cancer cell lines A549 (no-small cell lung cancer), N059 K (glioblastoma), and MCF-7 (breast cancer) by DCA (dichloroacetate), the membrane potentials measured by URFD were reversed to the low values (i.e. mitochondrial normal function was restored). Therefore, the high value of URFD does not indicate a large mitochondrial activity and very likely depends on the distribution of positive and negative ions in the cell (K^+, lactate) and/or the ordered water layer around mitochondria. Nevertheless, measurement of the mitochondrial membrane potential enables us to distinguish the two phenotypes of cancers.

The function of biological organs depends on mutual interactions and cooperation between the cells in the tissue. Generally, the long-range interactions depend on the generated electromagnetic fields, their frequency spectra and spatial patterns. The cancer cell may escape from interactions with the surrounding healthy cells and perform individual activity if its frequency spectrum of the electromagnetic field is rebuilt and shifted[7,54,55] and/or spatial pattern disturbed. The spatial pattern depends on the geometrical arrangement of the cytoskeleton structures, in particular of the microtubules and their excitation. Human nontumorigenic epithelial breast cells have a smaller deformability than the cells with increased metastatic potential — 10 and 30%, respectively.[78] The bioactive lipid SPC that influences the cancer metastasis causes shrinking of the keratin network around the nucleus[79] and consequently may cause diminished interactions between the cells based on the microtubule oscillations. It is not clear whether such keratin defects are connected with shrinking of the nuclear membrane (wrinkling) used as one of the important diagnostic markers in examination of gynecological cancers. The interaction forces between cancer cells may differ from those between healthy cells or between a healthy and a cancer cell. The force effect together with disturbances of the intercellular matrix may constitute an essential part of the local invasion and metastasis. This process is well described and referred to as the epithelial-to-mesenchymal transition.

Escape of a cell from the tissue subjection and its independent activity is a basis of the malignant properties. The cancer cells with the mitochondrial dysfunction have a lower biological activity than healthy cells. The cancer cells with fully functional mitochondria and a supply of energy rich metabolites display a higher biological activity than healthy cells and a high aggressiveness (but together with fibroblast very likely a lowered total energy of the coherent states far from the thermodynamic equilibrium). The effects of mitochondrial dysfunction and overfunction can be assessed on the basis of the electrodynamic fields generated by microtubules. The deviation of the power of the microtubule oscillations to lower or higher values results in corresponding frequency shifts and a loss of the interaction forces with the cells in the healthy tissue. The absorption resonant frequency of some cancer tissues were found at about 465 MHz.[23] The frequency of the microtubule oscillations depends on the nonlinear characteristic of the microtubule oscillators. If the force constant in the potential valley increases with the decreased excitation power then the frequency 465 MHz of the glycolytic phenotype cancer

Cancer transformation pathway

Initial phase (latent)	Intermediate phase (precancerous)	Final phase (cancer)

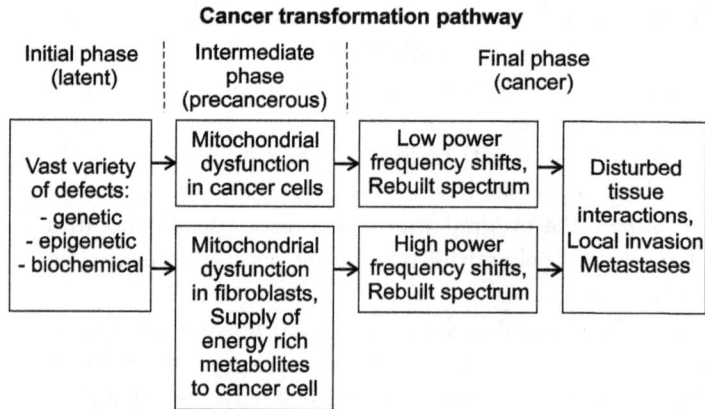

Fig. 1.   A scheme of the cancer transformation pathway. The pathway is divided in three main phases containing (a) chemical, genetic, and epigenetic disturbances (the initial phase), (b) formation of mitochondrial dysfunction in cancer cells or associated fibroblasts (the intermediate phase), and (c) changes of power and frequencies and consequent malignant deviations (the final phase).

corresponds to the shifted spectral lines of the healthy cells in the frequency band below 200 MHz. A schematic plot of the cancer transformation pathway (including frequency shifts) is shown in Fig. 1.

7. Discussion

Chemical, genetic, and physical mechanisms make up tools for biological functions. These mechanisms by themselves represent manifestation of a system whose living essentials are not yet clear. The chemical reactions and the physical processes of biological systems may exist in the inanimate world too but all of them together demonstrate the animate system. Cells subjected to the tissue ordering are very likely capable of acting like independent entities. In the tissue the cell performs mechanistic activities which can be described in terms of chemical signaling, transfer of information encoded in the genome for production of proteins, and other operations under general control executed by the tissue, the brain, and the rest of the body. The cell is an obedient machine which under control does not express its possible independence. However, the cell performs all its duties under internal instructions — very likely programs inscribed in internal structures. The nuclear DNA contains information for the material building in the protein coding part (which occupies a fraction smaller than 2%). The rest of the human genome forms the noncoding sequences. A few percent of the genome contains highly conserved parts of noncoding DNA which is an evidence of a strong evolutionary pressure. It may be assumed that the noncoding part of DNA contains a wide spectrum of information for chemical reactions and physical processes in the cells. The essential physical parameters for establishment of the coherent oscillations of cells in the tissue and the whole body are the power and the frequency of oscillations in the cellular structures. Fröhlich assumed the generating structures in the plasma membrane.[6] Del Giudice *et al.*[80] derived that the quantum self-focusing causes confinement of the electromagnetic waves to signals in a region whose dimension corresponds to the microtubule internal diameter. The idea of the coherent excitation in the cytoskeleton was analyzed by Hameroff[81] and Fröhlich included this contribution into the volume devoted to the biological coherence. Penrose[82] related consciousness to the action of the cytoskeleton and to microtubules in particular. Generation of Fröhlich's polar vibrations in microtubules was proposed by Pokorný *et al.*[30] The oscillations in the tubulin heterodimers in the microtubules interact with the ordered water and the free charges in the water inside the microtubule cavity that conditions the microtubule function.[32] The ordered water around microtubules provides low damping[49] and may also participate in the coherent oscillations. (The layers of the ordered water are formed by a strong static electric field, for instance around the charged surfaces of protein structures,[44] mitochondria,[50] etc.) Therefore, the physical properties of heterodimers and the level of their excitation belong to the essential parameters.

The frequency depends on the potential valleys of the heterodimer oscillators. The tubulin heterodimers encoded by the genome display a variety of conformation states (caused by dipolar arrangement) dependent on the electric parameters.[83] Different dipolar arrangements should result in different potential valleys. The measured spectra of the resonant frequencies of oscillations in the microtubules in the frequency band below 2 GHz suggest a nonlinear nature of the potential valleys which causes dependence of the frequency of oscillations on the power.[32] Besides that, the properties of the potential valleys may be changed by the static electric field created by mitochondria along the microtubules and the resonances depend also on the interaction of oscillations with the mobile charges in the ordered water filling in the microtubule cavity. The cell should maintain the frequency and the power of oscillations at the required values either through interaction with the other cells or by regulation. The regulation process needs frequency and power standards. The sequences of bases (adenine, etc.) might be inscribed into the noncoding part of DNA and after a convenient translation could represent the standards for regulation. DNA in the nucleus represents the ROM memory of the cell. A program of the cellular activity might be an issue of evolution, development, and storage in the memory. Mechanisms of its realization could depend on the microtubules which are engaged in a number of cellular activities. A single microtubule has a memory capacity of about 500 bits.[83] The total capacity of the digital memory of microtubules in a cell is about 20 kB. Therefore, microtubules could form a RAM memory. Microtubules have another remarkable property. They behave like biomolecular transistors capable of amplifying electric signals.[84] Microtubules with a centrosome could perform a role of a processing unit. Therefore, the cell seems to represent a functional system whose activity is programed and experience stored in the memory.

The malignant properties of cancer cells are created after significant disturbance of the oxidative metabolism. The power and the frequency of the microtubule oscillations are altered, which leads to a loss of interaction in the tissue and independent behavior of the cell. This pathological state may be caused by the defects of the coding and/or noncoding sequences of the genome. The former case leads to pathological production of some proteins, the latter case to defective mechanisms, functions, or setting some incorrect parameters (e.g. the frequency). Nevertheless, a parasitic consumption of the energy may also lead to changes of the energy level and loss of interaction of a cell with the tissue. Seyfried and Shelton[85] published a hypothesis that the mitochondrial dysfunction is a primary cause of cancer. The hypothesis is in contradiction with the observation that mitochondrial dysfunction is formed in the period of precancerous lesions developed after previous changes. On the other hand the Seyfried and Shelton hypothesis indirectly supports the idea that the lactate dehydrogenase elevating virus (LDV) is somehow connected with origin and/or development of the cancer process. LDV establishes lifelong persistent viremia in mice, parasites on the energy system, and is observed as a dark particle at the mitochondria.[12] The possible role of LDV in the cancer origin, course, and progress is not clear. LDH virus antigen elicits a response of the cell mediated immunity of T lymphocytes prepared from venous blood of the patients with malignant tumors (breast, gynecological, laryngeal and pharyngeal).

The microtubule oscillations and the electromagnetic field might be disturbed by increased conductivity in the cell. Asbestos cancerogenicity (mainly production of mesothelioma) is explained by the capability of the asbestos fibres to short-circuit distant parts of the cell with different levels of the electromagnetic field. Asbestos may form optic fibres for the cellular field[86] and/or conductive wires after adsorption of specific proteins and molecules containing iron atoms at the fibres surface.[87]

The high values of URFD measured at the mitochondrial inner membrane in cancer cells with the normal Warburg effect are explained by hyperpolarization. It contradicts the fact that mitochondria are dysfunctional. The real membrane potential corresponding to the proton transfer is smaller than in a healthy cell. The value of the measured potential rather corresponds to the increased amount of the lactate produced as a result of inhibition of the pyruvate transfer into the matrix and reaction of the lactate dehydrogenase enzyme. The production of negatively charged lactates might in a final effect result in an increased number of hydrogen ions to maintain electroneutrality. It should also be mentioned that the experimental data from URFD measurement were interpreted without analysis of the effect of the ordered water layer around mitochondria. The water ordering is not significant if the membrane potential and the intensity of the electric field are low. All transferred protons are concentrated at the membrane. For higher membrane potential an ordered layer of water is built. The protons may be distributed in two layers. The inner layer is at the membrane, the outer layer at the outer rim of the ordered water. Therefore, only a part of the membrane potential is measured.

This might be the case of the healthy cells and the cancer cells with the reverse Warburg effect.

Measurement of the electrodynamic activity of living cells is a challenge for nanotechnology. The power of the oscillating field is extremely low. The whole cell processes a power of about 0.1 pW.[88] A part of the power may be transformed into the polar modes in individual microtubules. Each of about 400 microtubules in a cell might be excited by a power of the order of magnitude 0.1 fW. The quality factor of a microtubule is very likely not much less than 100 (determined from data of Sahu *et al.*[32]) and the total power which may be possibly released from oscillations in the microtubule cellular structure should correspond to about 10 fW. Microtubules in the cell are assumed to be arranged in a symmetrical system (for instance discrete spherical symmetry in a spherical cell) and excitation of the oscillations should correspond to the same symmetry. In this case the emission losses are low. If the power measured at the plasma membrane depends only on the electrodynamic oscillations of a single microtubule, a power of the order of magnitude of 0.1 fW or lower could be detected. Another requirement concerns the dimension of the sensor contact. The electrodynamic field of an individual microtubule bound to a structure at the membrane could be measured at a region with the linear dimensions smaller than about 100 nm. Measurement should be provided at a "biological" temperature.

The mitochondrial dysfunction is a key point in the cancer development. It is a point of transition to malignancy. A large variety of the genetic, epigenetic, and biochemical disturbances result in the dysfunction of mitochondria in cancer cells or in their associated fibroblasts. Both phenotypes of cancer cells (i.e. cancer cell with dysfunctional and fully functional mitochondria) are protected against apoptosis. Treatment of cancers at the link of the mitochondrial dysfunction seems to be promising. A normal function of mitochondria in cancer cells with the Warburg effect could be restored by blocking PDKs (or their production) inhibiting the pyruvate transfer into the mitochondrial matrix. A normal function of fibroblasts in cancers with the reverse Warburg effect could be achieved by cutting off the pathological signaling from the cancer cell to fibroblasts, transport of the energy rich metabolites from fibroblasts to the cancer cell, and restoring a normal mitochondrial function in fibroblasts. After restoration of a healthy state the cells do not necessarily have to continue in their normal function. A restored cancer cell which has been damaged to a large extent would enter the apoptotic process. However, the cancer cells of both phenotypes could also return to the pathological cancer state. It may signify that the conditions for creation of the pathological state remain unchanged and that they might be stored in ROM and/or RAM memory of the cell. But even temporary transition of cells from the cancer to the healthy state could make possible an effective treatment of cancers of the fourth stage. In general, targeting mitochondrial link of the cancer transformation can represent a big step forward in the cancer treatment.

8. Conclusion

One of the main differences between the animate and inanimate systems is a coherent state far from the thermodynamic equilibrium dependent on the energy supply. In multicellular bodies this coherent nonequilibrium state depends on the oxidative metabolism of mitochondria which transforms the chemical energy into a convenient form and adjusts the conditions for excitation of the coherent polar oscillations in microtubules. Mitochondria — an almost universal part of the eucaryotic cells — are multifunctional organelles which in the process of energy transformation produce layers of a strong static electric field and ordered water around them. These layers make possible low damping of the polar oscillations in microtubules and their shift into a highly nonlinear region. The energy stored in the polar oscillations in microtubules forms a low frequency component of the coherent state far from the thermodynamic equilibrium of the biological system. Microtubules are one of the oscillating structures generating the cellular electromagnetic field whose function includes participation in the interactions with the surrounding cells in the tissue. Excitation of polar oscillations is crucial for multicellular organisms.

Disturbances of the mitochondrial function, the cytoskeleton structure, and the oscillations in microtubules cause defects of the coherent state far from the thermodynamic equilibrium which endangers life. An important link of the cancer transformation pathway contains mitochondrial dysfunction in the cancer cells (the normal Warburg effect) or in the fibroblasts associated with the cancer cells (the reverse Warburg effect). This transformation produces inhibition of the apoptotic function so that the cancer cells are immortal. The mitochondrial dysfunction resulting in disturbances of the electromagnetic field leads to an independent behavior of the cell in the multicellular body. Malignancy is created. Cancer is a pathology of the coherent state far from the thermodynamic equilibrium. The treatment should be primarily targeted to restore the mitochondrial function.

The cells subjected to the tissue seem to be able to perform activity of an independent living entity. Instead, the cell carries out mechanistic work under general control of the tissue assembly of cells, the brain, and the rest of the body. The cell is a tissue-abiding machine. But all its duties seem to be performed under internal programs stored in the genome in its coding and noncoding parts. DNA in the nucleus represents a ROM of the cell. The microtubules can store information and may be considered as a RAM type memory. The microtubules are also capable of providing a role of the processing unit. A program for the cell functions and processes is stored in the memory and very likely could be changed on the basis of the cell experience or random events. Cancer cell could be also governed by a changed memory program.

Acknowledgments

This study was supported by grant No. P102/11/0649 of the Czech Science Foundation GA CR.

References

1. A. Jandová, K. Motyčka, J. Čoupek *et al.*, *Sborník lékařský* (in Czech) **81**, 321–327 (1979).
2. H. Fröhlich, in *Theoretical Physics and Biology*, ed. M. Marois (North Holland, Amsterdam, 1969), pp. 13–22. (*Proc. 1st Int. Conf. Theor. Phys. Biol., Versailles, 1967*).
3. H. Fröhlich, *Phys. Lett. A* **26**, 402–403 (1968).
4. H. Fröhlich, *Int. J. Quantum Chem.* **II**, 641–649 (1968).
5. H. Fröhlich, *J. Collect. Phenom.* **1**, 101–109 (1973).
6. H. Fröhlich, *Adv. Electronics Electron. Phys.* **53**, 85–152 (1980).
7. H. Fröhlich, *IEEE Trans. MTT* **26**, 613–617 (1978).
8. O. Warburg, K. Posener and E. Negelein, *Biochem. Z.* **152**, 309–344 (1924).
9. O. Warburg, *Science* **123**, 309–314 (1956).
10. S. Bonnet, S. L. Archer, J. Allalunis-Turner *et al.*, *Cancer Cell* **11**, 37–51 (2007).
11. S. Pavlides, D. Whitaker-Menezes, R. Castello-Cros *et al.*, *Cell Cycle* **8**, 3984–4001 (2009).
12. A. Jandová, J. Pokorný, J. Kobilková *et al.*, *Electromagn. Biol. Med.* **28**, 1–14 (2009).
13. H. A. Pohl, *Int. J. Quantum Chem. Quantum Biol. Symp.* **7**, 411–431 (1980).
14. H. A. Pohl, T. Braden, S. Robinson, J. Piclardi and D. G. Pohl, *J. Biol. Phys.* **9**, 133–154 (1981).
15. S. C. Roy, T. Braden and H. A. Pohl, *Phys. Lett.* **83A**, 142–143 (1981).
16. H. A. Pohl, *Dielectrophoresis* (Cambridge Univ. Press, London, 1978).
17. R. Hölzel and I. Lamprecht, *Neural Netw. World* **4**, 327–337 (1994).
18. R. Hölzel, *Electro Magnetobiol.* **20**, 1–13 (2001).
19. J. Pokorný, J. Hašek, F. Jelínek, J. Šaroch and B. Palán, *Electro Magnetobiol.* **20**, 371–396 (2001).
20. A. E. Pelling, S. Sehati, E. B. Gralla, J. S. Valentine and J. K. Gimzewski, *Science* **305**, 1147–1150 (2004).
21. A. E. Pelling, S. Sehati, E. B. Gralla and J. K. Gimzewski, *Nanomedicine: Nanotechnology, Biology, and Medicine* **1**, 178–183 (2005).
22. J. Pokorný, J. Hašek, J. Vaniš and F. Jelínek, *Indian J. Exper. Biol.* **46**, 310–321 (2008).
23. C. Vedruccio and A. Meessen, in *Proceedings PIERS, Progress in Electromagnetics Research Symposium*, Italy, Pisa, March 28–31, 2004, pp. 909–912.
24. G. Albrecht-Buehler, *J. Cell Biol.* **114**, 493–502 (1991).
25. G. Albrecht-Buehler, *Proc. Natl. Acad. Sci. USA* **89**, 8288–8293 (1992).
26. G. Albrecht-Buehler, *Proc. Natl. Acad. Sci. USA* **102**, 5050–5055 (2005).
27. F.-A. Popp, in *Herbert Fröhlich, FRS. A Physicist Ahead of His Time*, eds. G. J. Hyland and P. Rowlands (The University of Liverpool, Liverpool, 2006), pp. 155–192.
28. L. A. Amos and A. Klug, *J. Cell. Sci.* **14**, 523–549 (1974).
29. J. A. Tuszyński, S. Hameroff, M. Satarić, B. Trpisová and M. L. A. Nip, *J. theor. Biol.* **174**, 371–380 (1995).
30. J. Pokorný, F. Jelínek, V. Trkal, I. Lamprecht and R. Hölzel, *J. Biol. Phys.* **23**, 171–179 (1997).
31. E. D. Kirson, Z. Gurvich, R. Schneiderman *et al.*, *Cancer Res.* **64**, 3288–3295 (2004).
32. S. Sahu, S. Ghosh, B. Ghosh *et al.*, *Biosens. Bioelectron.* **47**, 141–148 (2013).
33. J. Pokorný, *Bioelectrochem.* **63**, 321–326 (2004).
34. J. Pokorný, J. Pokorný and J. Kobilková, *Integrative Biology* **5**, 1439–1446 (2013).
35. J. Pokorný, *Electromagn. Biol. Med.* **28**, 105–123 (2009).
36. F. Jelínek and J. Pokorný, *Electro Magnetobiol.* **20**, 75–80 (2001).
37. K. M. Tyner, R. Kopelman and M. A. Philbert, *Biophys. J.* **93**, 1163–1174 (2007).

38. L. A. Amos, in *Microtubules*, eds. K. Roberts and J. S. Hyam (Academic Press, London, New York, 1979), pp. 1–64.
39. H. Stebbings and C. Hunt, *Cell Tissue Res.* **227**, 609–617 (1982).
40. J. Zheng, W. Chin, E. Khijniak, E. Khijniak Jr. and G. H. Pollack, *Adv. Colloid Interface Sci.* **127**, 19–27 (2006).
41. G. Pollack, I. Cameron and D. Wheatley, *Water and the Cell* (Springer, Dodrecht, 2006).
42. B. Chai, J. Zheng, Q. Zhao and G. Pollack, *J. Phys. Chem. A* **112**, 2242–2247 (2008).
43. B. Chai, H. Yoo and G. Pollack, *J. Phys. Chem. B* **113**, 13953–13958 (2009).
44. G. H. Pollack, *The Fourth Phase of Water* (Ebner & Sons Publishers, Seatle, WA, USA, 2013).
45. E. C. Fuchs, J. Woisetschlager, K. Gatterer *et al.*, *J. Phys. D Appl. Phys.* **40**, 6112–6114 (2007).
46. E. C. Fuchs, K. Gatterer, G. Holler and J. Woisetschlager, *J. Phys. D Appl. Phys.* **41**, 185502-1–5 (2008).
47. E. C. Fuchs, B. Bitschnau, J. Woisetschlager *et al.*, *J. Phys. D Appl. Phys.* **42**, 065502-1–4 (2009).
48. L. Giuliani, E. D'Emilia, A. Lisi *et al.*, *Neural Netw. World* **19**, 393–398 (2009).
49. J. Pokorný, C. Vedruccio, M. Cifra and O. Kučera, *Eur. Biophys. J.* **40**, 747–759 (2011).
50. J. Pokorný, *AIP Adv.* **2**, 011207-1–11 (2012).
51. G. Preparata, *QED Coherence in Matter* (World Scientific, Hong Kong, New Jersey, London, 1995).
52. E. Del Giudice and A. Tedeschi, *Electromagn. Biol. Med.* **28**, 46–52 (2009).
53. J. Pokorný, *Electromagn. Biol. Med.* **22**, 15–29 (2003).
54. J. Pokorný and T.-M. Wu, *Biophysical Aspects of Coherence and Biological Order* (Prague: Academia; Berlin, Heidelberg, New York: Springer-Verlag, 1998).
55. J. Pokorný, in *Herbert Fröhlich, FRS. A Physicist Ahead of His Time*, eds. G. J. Hyland and P. Rowlands (The University of Liverpool, Liverpool, 2006), pp. 193–224.
56. J. Pokorný, *Electro Magnetobiol.* **20**, 59–73 (2001).
57. J. Pokorný, J. Hašek and F. Jelínek, *Electromagn. Biol. Med.* **24**, 185–197 (2005).
58. J. Pokorný, T. Martan and A. Foletti, *J. Acupunct. Meridian Stud.* **5**, 34–41 (2012).
59. D. E. Jaalouk and J. Lammerding, *Nat. Rev. Mol. Cell Biol.* **10**, 63–73 (2009).
60. D. E. Ingber, *Semin. Cancer Biol.* **18**, 356–364 (2008).
61. T. McFate, A. Mohyeldin, H. Lu *et al.*, *J. Biol. Chem.* **283**, 22700–22708 (2008).
62. R. Damadian, *Science* **171**, 1151–1153 (1971).
63. I.-Ch. Kiricuta, Jr. and V. Simplăceanu, *Cancer Res.* **35**, 1164–1167 (1975).
64. G. Bonuccelli, A. Tsirigos, D. Whitaker-Menezes *et al.*, *Cell Cycle* **9**, 3506–3514 (2010).
65. G. Bonuccelli, D. Whitaker-Menezes, R. Castello-Cros *et al.*, *Cell Cycle* **9**, 1960–1971 (2010).
66. B. Chiavarina, D. Whitaker-Menezes, G. Migneco *et al.*, *Cell Cycle* **9**, 3534–3551 (2010).
67. M. P. Lisanti, U. E. Martinez-Outschoorn, B. Chiavarina *et al.*, *Cancer Biol. Ther.* **10**, 537–542 (2010).
68. U. E. Martinez-Outschoorn, R. M. Balliet, D. B. Rivadeneira *et al.*, *Cell Cycle* **9**, 3256–3276 (2010).
69. U. E. Martinez-Outschoorn, C. Trimmer, Z. Lin *et al.*, *Cell Cycle* **9**, 3515–3533 (2010).
70. G. Migneco, D. Whitaker-Menezes, B. Chiavarina *et al.*, *Cell Cycle* **9**, 2412–2422 (2010).

71. S. Pavlides, A. Tsirigos, G. Migneco *et al.*, *Cell Cycle* **9**, 3485–3505 (2010).
72. S. Pavlides, A. Tsirigos, I. Vera *et al.*, *Cell Cycle* **9**, 2201–2219 (2010).
73. Y.-H. Ko, Z. Lin, N. Flomenberg *et al.*, *Cancer Biol. Ther.* **12**, 1085–1097 (2011).
74. U. E. Martinez-Outschoorn, Z. Lin, Y.-H. Ko *et al.*, *Cell Cycle* **10**, 2521–2528 (2011).
75. L. B. Chen, *Ann. Rev. Cell Biol.* **4**, 155–181 (1988).
76. E. D. Michelakis, L. Webster and J. R. Mackey, *Br. J. Cancer* **99**, 989–994 (2008).
77. J. S. Modica-Napolitano and J. R. Aprille, *Cancer Res.* **47**, 4361–4365 (1987).
78. J. Guck, S. Schinkinger, B. Lincoln *et al.*, *Biophys. J.* **88**, 3689–3698 (2005).
79. M. Beil, A. Micoulet, G. von Wichert *et al.*, *Nat. Cell Biol.* **5**, 803–811 (2003).
80. E. Del Giudice, S. Doglia and M. Milani, in *Coherent Excitation in Biological Systems*, eds. H. Fröhlich and F. Kremer (Springer-Verlag, Berlin, 1983), pp. 123–127.
81. S. R. Hameroff, in *Biological Coherence and Response to External Stimuli*, ed. H. Fröhlich (Springer-Verlag, Berlin, Heidelberg, New York, 1988), pp. 242–263.
82. R. Penrose, *Shadows of the Mind* (Oxford University Press, Oxford, New York, 1994).
83. S. Sahu, S. Ghosh, K. Hirata, D. Fujita and A. Bandyopadhyay, *Appl. Phys. Letters* **102**, 123701-1–4 (2013).
84. A. Priel, A. J. Ramos, J. A. Tuszyński and H. F. Cantiello, *Biophys. J.* **90**, 4639–4643 (2006).
85. T. N. Seyfried and L. M. Shelton, *Nutr. Metab. (Lond)* **7**, 7-1–22 (2011).
86. R. R. Traill, 9th Int. Fröhlich's Symposium, *J. Phys. Conference Series* **329**, 012017-1–15 (2011).
87. S. Toyokuni, *Nagoya J. Med. Sci.* **71**, 1–10 (2009).
88. I. Lamprecht, in *Biological Microcalorimetry*, ed. A. E. Beezer (Academic Press, London, 1980), pp. 43–112.

Chapter 12

Potential Mechanisms of Cancer Prevention by Weight Control

Yu Jiang* and Weiqun Wang[†]

Department of Human Nutrition, Kansas State University
Manhattan, KS 66506, USA
*yjiang@ksu.edu
[†]wwang@ksu.edu

Weight control via dietary caloric restriction and/or physical activity has been demonstrated in animal models for cancer prevention. However, the underlying mechanisms are not fully understood. Body weight loss due to negative energy balance significantly reduces some metabolic growth factors and endocrinal hormones such as IGF-1, leptin, and adiponectin, but enhances glucocorticoids, that may be associated with anti-cancer mechanisms. In this review, we summarized the recent studies related to weight control and growth factors. The potential molecular targets focused on those growth factors- and hormones-dependent cellular signaling pathways are further discussed. It appears that multiple factors and multiple signaling cascades, especially for Ras-MAPK-proliferation and PI3K-Akt-anti-apoptosis, could be involved in response to weight change by dietary calorie restriction and/or exercise training. Considering prevalence of obesity or overweight that becomes apparent over the world, understanding the underlying mechanisms among weight control, endocrine change and cancer risk is critically important. Future studies using "-omics" technologies will be warrant for a broader and deeper mechanistic information regarding cancer prevention by weight control.

1. Introduction

Obesity rate in the U.S. is growing rapidly during the past 20 years.[14] It has become a serious worldwide problem which is associated with increased risk for several chronic diseases, including cancer, diabetes, and cardiovascular disease. Studies showed evidence for a positive association between overweight/adiposity and cancer risk in esophagus, pancreas, colon, rectum, endometrium, kidney, and postmenopausal breast cancer.[a] Weight control, therefore, has become an important strategy against cancer and/or other chronic diseases. Body weight control is carried out by the balance of negative energy, which is tightly associated with dietary calorie intake and/or physical activity (energy expenditure). A positive energy balance, via increased dietary intake and/or decreased energy expenditure, results in

---

[a]WCRF/AICR, http://www.dietandcancerreport.org/?p=ER

increased weight and fat mass, or adiposity. Negative energy balance via decreased calorie intake or increased expenditure in adult may help maintain body weight and thus benefit health status.

Calorie restriction is referred to as decrease of energy intake without malnutrition. In calorie restriction regimens, proteins and all the essential micronutrients such as vitamins and minerals are kept same. Only the total amount of energy from fat and carbohydrate is reduced usually at 20–40% of the *ad libitum*-fed controls. The cancer preventive effect of calorie restriction has been found for almost 100 years. The first animal study was done as early as 1909 by Moreschi, who observed that tumors transplanted into underfed mice grew slower than those in *ad libitum*-fed mice.[75] In the 1940s, Tannenbaum and colleagues found that reduced food intake decreased tumor incidence in experimental animals.[106] Later on, the preventive effect of calorie restriction on cancer is confirmed in various animal models such as primate and rodent or various organs including mammary gland, prostate, colon, and skin. Calorie restriction has been shown to be effective in both spontaneously occurring and chemically induced cancers. Calorie restriction is also able to lessen cancer in genetically engineered models, e.g. p53 knock out mice and APC[min] mice.[47,69] To date, calorie restriction is found to be the most potent and effective dietary intervention strategy for cancer prevention in animal models.[46]

The health benefit of physical activity (exercise), on the other hand, has been known for many decades. Accumulated evidence both in human studies and animal models has shown that physical activity is helpful in decreasing cancer risk. The epidemiologic studies on the relationship between physical activity and cancer prevention as reviewed by Friedenreich and Orensterin[32] suggested that the evidence of cancer prevention by physical activity be convincing for colon and breast cancer, probable for prostate cancer, and possible for endometrium and lung cancer, although some other types of cancers seemed less sufficient and conclusive. Colon cancer is studied most with respect to physical activity in animal models. It was found that physical activity, both by forced treadmill and voluntary wheel, was effective in reducing azoxymethane-induced colon carcinomas in rats.[68,110] However, the results were not conclusive in APC[min] mice.[5] The effect of physical activity on breast cancer prevention in animal models was reviewed by Thompson *et al.*,[108,109] indicating physical activity might inhibit mammary carcinogenesis, but the effect was less reproducible compared to calorie restriction. Overall, the impact of physical activity on cancer prevention is positive, but not consistent or potent as calorie restriction approach.

Despite many studies have been conducted, no mechanism of weight control on cancer prevention has been well-established. Enhancement of DNA repair and diminution of oxidative damage to DNA, as well as reduction of oncogene expression have been postulated. Weight loss, via calorie restriction and/or exercise, has been found to reduce certain circulating growth factors and hormones, such as IGF-1 and adipocytokins, but enhance glucocorticosteroids, which are critical in maintenance of cellular growth, proliferation, cell cycle, and apoptosis function. Reduction of these growth factors and inhibition of these factors-dependent biological processes by weight control may contribute to the overall anti-carcinogenesis.

2. IGF-1: A Key Modulator for Cell Growth and Anti-Apoptosis

2.1. *IGF-1 system: IGF-1, IGF-1 binding proteins, IGF-1 receptor and IGF signaling*

Insulin-like growth factors (IGF-1 and IGF-2) are 70-amino-acid polypeptides that have high sequence similarity to insulin. Both IGF-1 and IGF-2 have metabolic functions, and play important roles in cellular proliferation and differentiation. The major function of IGF-2 seems related to embryonic growth and early development,[22] but IGF-1 is more important in post-natal growth. The synthesis of IGF-1 is mainly regulated by growth hormone in liver. IGF-1 is majority produced in the liver but also all the other cells.

The circulating levels of IGF-1 and their bioavailability are modulated by a family of IGF binding proteins (IGF-BPs), which have six homologies. IGF-BP3 is the most abundant in humans. It is found that about 90% of IGF-1 in the serum is binding to IGFBP3, a complex formed that is very large and cannot transport out of bloodstream. Free IGF-1 or IGF-1 that binds to IGF-BP1 and IGF-BP2 are able to across the capillary endothelium and reach target tissues.[120] IGF-BPs are degraded by proteases both in the tissues and in the circulation, through which IGF-1 is freed and interacts with IGF-1 receptors. Furthermore, IGF-BPs can also modulate the process of IGF-1 binding to receptors and the IGF signaling.[42,19]

IGF-1 receptor (IGF-1R) is a member of receptor tyrosine kinase super-family. IGF-1R binds IGF-1 with the highest affinity, while it can also bind IGF-2 and insulin.[116] In addition to IGF-1R, insulin receptor (IR) and IGF-2 receptor (IGF-2R) are able to bind IGF-1, but with less affinity. Studies showed that insulin receptor and IGF-1 receptor could form heterohybrid.[98,81]

Binding of IGF-1 to the receptors will induce autophosoporylation and activation of downstream signal network, such as phosphatidylinositol-3-kinase (PI3K) pathway. The phosphorylation of PI3K results in activation of Akt. Activated Akt will then inhibit the activation of interleukin-1β-converting enzyme (ICE)-like protease, therefore suppresses apoptosis. Binding of IGF-1 to its receptor is also found to activate other pathways, such as MAPK pathway. Activation of MAP kinase will lead to increase of cell proliferation.[136]

2.2. *IGF-1 and cancer*

The IGF-1 system has been found to be involved in human development and the maintenance of a normal function and homeostasis of the cell growth in the body. Abnormal function in increased IGF-1 levels leads to break down of normal cell homeostasis and function, which are usually found in cancer development. Studies showed that the neoplasia process may be due to the elevation of IGF-1 in the circulation and/or the increased sensitivity of IGF-1R to the hormone. Increased IGF-1 stimulates cell proliferation and inhibited apoptosis in various cancer cells. The relationship of IGF-1 and cancer and the potential corresponding mechanism have also been studied extensively in human subjects and animal models.

In the study of colon cancer, it was found that the gene expression of IGF-1 was elevated in colon carcinomas.[113] Similar results were also found in the human breast and lung tumors.[130] Later on epidemiological studies showed plasma IGF-1 levels were positively associated with higher risk of cancer, especially prostate and breast cancer. Chan *et al.*[15] first demonstrated a link between circulating IGF-1 and prostate cancer risk using a nested case-control study. The study showed that plasma IGF-1 levels were positively associated with prostate cancer risk. Comparing to men in the lowest quartile of plasma IGF-1 levels, men in the highest quartile had a 4.3 folder higher risk of prostate cancer.[15] In women, IGF-1 was found to be positively associated with pre-menopausal breast cancer rather than post-menopausal breast cancer.[89,120,11] The correlation between circulating IGF-1 levels and cancer risk was also found in colon cancer and bladder cancer.[36,133] In addition to cancer risk, elevated plasma IGF-1 was associated with benign prostatic hyperplasia, proliferation of colorectal mucosa, and colorectal adenomas.[59,18,13,89,107] Overall, the reported studies support that relatively high circulating IGF-1 levels may have a causal role in cancer development.

IGF-1 and IGF-1 signaling are also found to play an important role in skin cancer. Rho *et al.*[90] found that the mRNA level of IGF-1 and IGF-1 receptor in dermal and epidermal of mouse skin was significantly increased in the skin papillomas and carcinomas. In order to detect the potential role of IGF-1 signaling in the multistage mouse skin carcinogenesis, DiGiovanni lab developed transgenic mice HK1.IGF-1, in which IGF-1 is over expressed in epidermis driven by a human keratin 1 promoter.[8] The authors found that HK1.IGF-1 transgenic mice were more sensitive to tumor promoters such as TPA, chrysarobin, okadaic acid, and benzyl peroxide after initiated by DMBA than wild type mice. Comparing to wild type animals which received the same dose of carcinogen treatment, transgenic mice developed tumors more rapidly and the number of tumors per mouse were dramatically increased.[8,124] In addition, squamous papillomas and carcinomas were found to develop spontaneously in a similar transgenic mouse model BK5.IGF-1, which over expresses IGF-1 in the basal layer of skin epidermis.[24] Activation of IGF-1 receptor, epidermal hyperplasia and increased labeling index were also observed in these mice. Not only in chemically induced skin carcinogenesis, an altered IGF system were also found to contribute to HaCaT keratinocyte UV susceptibility.[111] The above data suggested that constitutive expression of IGF-1 and activation of IGF-1 receptor signaling pathways in basal epithelial cells lead to tumor promotion, in which IGF-1 played an important role in skin cancer development. More recently, it was found that PI3K/Akt pathway is important in IGF-1 mediated skin promotion.[125] Inhibition of PI3K activity significantly blocked epidermal proliferation, as well as skin tumor development in DMBA initiated IGF-1 transgenic mice.[125]

2.3. *IGF-1 as a mediator in cancer prevention by weight control*

As discussed above, the high IGF-1 levels seem associated with the risk of cancer development and lowering IGF-1 levels via weight control appears to be related to a decreased cancer incidence. Thus, manipulating plasma IGF-1 levels have been applied in cancer prevention strategies. In order to test this hypothesis, a mouse model that has a genetic deletion of liver IGF-1 gene was generated.[126] In these mice, IGF-1 levels are 25% of that in nontransgenic mice. Lowering circulating IGF-1 significantly delayed mammary gland tumor development by carcinogen DMBA or C3 (1)/SV40 large T-antigen induced carcinogenesis.[126] Fibroblasts lacking IGF-1 receptor were found to be highly resistant to transformation by simian virus 40 T antigen.[85,95] Moore *et al.*[74] found that the activation of the Akt and mTOR signaling pathways by tumor promoter TPA were significantly reduced in IGF-1 deficient mice, resulting in a blockage of epidermal response to tumor promotion. Kari *et al.*[56] found that the functional disruption of IGF-1R markedly inhibited breast cancer metastasis in the nude mice by suppressing cellular adhesion, invasion, and metastasis of breast cancer cells to the lung, lymph nodes, and lymph vessels.

Reducing plasma IGF-1 by weight control has been investigated in a number of studies. Ruggeri *et al.*[93] first reported that dietary calorie restriction decreased serum IGF-1 significantly at the first and third week after the experiment started in female Sprague-Dawley rats. Hursting *et al.*[48] found serum IGF-1 in 40% of calorie restricted rats was only 44% of *ad libitum*-fed controls. The author also infused human recombinant IGF-1 back to the dietary restricted rats by using osmotic minipumps. Infusion of IGF-1 restored cell proliferation activity and enhanced mitogen responsiveness in dietary restriction treated rats.[48] In a tumor study of p53 deficient mice, 20% of calorie restriction decreased circulating IGF-1 by 26% and restoration of IGF-1 in calorie restricted mice did not change the tumor incidence significantly, but increased cell proliferation and inhibited apoptosis dramatically.[25] Study by Thompson also found that 40% of calorie restriction reduced circulating IGF-1 by half in rats and restoration IGF-1 failed to have effect on mammary tumor incidence.[135] Studies in our lab also found that IGF-1 was significantly decreased by dietary calorie restriction and restoration of IGF-1 significantly abolished PI3K reduction in treadmill exercised mice with limited feeding at same amount as sedentary control.[127] Overall, the above results showed that reduction of IGF-1 levels and thus down-regulation of IGF-1 signaling pathways as a consequence of dietary restriction could contribute to anti-tumorigenesis. Restoration of IGF-1 abrogated, at least in part, the protective effect of calorie restriction on carcinogenesis.

The impact of physical activity on IGF-1 reduction and cancer prevention is complicated. As reviewed by Kaaks *et al.*,[53] physical activity decreased IGF-1 level in children and adolescents. But for adults, the plasma IGF-1 levels were not decreased either by short bout exercise or physical training. Studies showed that weight control by long-term exercise could decrease IGF-1. For example, a recent published paper found that plasma concentrations of IGF-1 were significantly lower in endurance runners than sedative controls.[29] In animal model, our lab found that exercise alone with *ad libitum* feeding was not sufficient to decrease plasma IGF-1 levels. When the exercised mice was fed with the same amount as their sedentary counterpart, plasma levels of IGF-1 were modestly but significantly reduced (unpublished data). Nevertheless, the evidence by ours and others indicate that a negative energy balance appears to be a fundamental requirement for IGF-1 reduction and potential cancer prevention.

3. Adipocytokines: A Linkage of Adipose and Cancer Risk

Adipocytokines are secretary products of adipose tissue and have metabolic and endocrine functions. They include leptin, adiponectin, resistin, and visfatin, etc., which have been identified and studied recently for a potential relationship between obesity and cancer risk.[137]

3.1. *Leptin*

Leptin gene, which is also called obese (ob) gene, encodes a 16 kDa protein.[132] As an adipocytokine, leptin is secreted mainly by adipose tissue. Other tissues, such as placenta, ovaries, skeletal muscle, pituitary gland, stomach, and liver, are also able to produce leptin. The major factor that affects circulating leptin levels is adipose tissue mass.[68] Increased body weight has been shown to be positively associated with high level of plasma leptin.[31] Leptin was found to regulate appetite and control body weight through affecting the hypothalamus, suppressing food intake and stimulating energy expenditure.[76] In addition to the central circuits, leptin also has effects in the periphery tissues, such as lung, intestine, skin, stomach, heart and other organs, though binding to leptin receptors.[70,20]

Leptin receptors contain extracellular, transmembrane, and intracellular domains. The extracellular domain is responsible for leptin binding and intracellular domain recruits and activates downstream substrates. Activation of leptin receptors was found to stimulate signaling pathways, such as JAK2/STAT3, Ras/ERK1/2, and PI3K/Akt/GSK3. Other signaling proteins induced by leptin were also found, including protein kinase C, p38 kinase, and AP-1 component c-fos, c-jun, and junB, etc. (reviewed by Garofalo *et al.*[33]).

Leptin is important in the regulation of energy balance. Obese (ob/ob) mice, which have leptin gene mutation, are found to be morbidly obese, infertile, hyperphagic, hypothermic, and diabetic.[44] Infusion of recombinant leptin into these mice reduced food intake and decreased body weight.[10,39] In diet induced obese mice, the circulating leptin was significantly elevated with the increase of body weight. Studies also showed that these mice are resistant to peripherally administrated leptin.[118] Compared to normal weight people, obese people usually developed hyperleptinemia and leptin resistant, which might be due to a defect in transporting of leptin through the blood barrier.[3,12]

Epidemiologic studies showed that moderately elevated serum leptin was associated with prostate cancer development.[100] People that have high leptin levels tend to have a large tumor.[16,94] However, some studies found that there was no relationship between circulating leptin and prostate cancer risk.[63,99] In vitro, leptin is found to be a promoter in cancer cells. Studies showed that leptin induced cell proliferation in breast cancer ZR75-1 and HTB-26 cells via the activation MAPK and PI3K.[30] Leptin also simulated estrogen synthesis by increasing aromatase gene transcription and protein activity, which implied that leptin might be responsible for the resistance to anti-estrogens during hormonal treatment of breast cancer.[104] In colon cancer cells, leptin induced cell growth and blocked apoptosis of human cancer HT29 cells via stimulation of ERK1/2 and NFκB pathway.[40,65] In addition, the mitogentic activity of leptin has also been demonstrated in prostate, pancreatic, ovarian, and lung cancer cells. Taken together, leptin seems to be important in tumor progression. Manipulation of plasma leptin might be effective in cancer prevention and treatment.

As discussed above, leptin was positively associated with body weight and body mess index. Weight control seems to be effective in lowering circulating leptin. Fontana *et al.*[29] showed that plasma concentrations of leptin were significantly lower in endurance runners than sedative controls. In animal models, it was found that 40% of calorie restriction significantly decreased the serum leptin levels in APC^{min} mice compared to the control.[69] Studies by our laboratory showed that the plasma level of leptin was significantly decreased in calorie restricted mice and exercised mice with paired feeding, but not in exercised mice with *ad libitum* feeding.[127] Interestingly, we found that leptin in subcutaneous fat cells was not affected by weight control treatment.[127] All the above evidence suggests that leptin can be important in medicating the cancer protective effects of weight control. Further research is needed to characterize the specific role of leptin in cancer development.

3.2. *Adiponectin*

Adiponectin is also an adipocytokine that is secreted in adipose tissue and plays an important role in obesity-related disorders. The gene of adiponectin is located on diabetes susceptibility locus chromosome 3q27.[97,105] Adiponectin was found to account for 0.01% of total plasma protein in human serum.[2] It exists in several forms: trimers, hexamers, high molecular weight multimers (HMW), or globular form. HMW form was suspected to be the most bioactive form.[91,28]

Two adiponectin receptors have been identified. The signaling downstream of adiponectin receptors is still under investigation. Miyazaki *et al.*[72] found that different forms of adiponectin have distinct biological effects, which may be through differential activation of downstream signaling.

Some evidence has showed that adiponectin is an insulin sensitizing hormone and may process anti-diabetic activities.[55] The blood level of adiponectin was found to be lower in obese people and was able to neutralization of LPS activity and anti-inflammation.[115,119] In addition, adiponectin was also found to be a modulator of lipid metabolism and might have preventive effect on cardiovascular disease. Kim et al.[60] found that an increase of adiponectin concentrations or the maintenance of the higher levels was negatively associated with cardiovascular risk factors in nondiabetic CAD male patients, independent of adiposity and smoking status.[60]

The potential anticancer properties of adiponectin have been investigated both in epidemiological study and animal models. There are three case control studies which showed that low serum adiponectin levels were associated with an increase risk of breast cancer in women.[73,17,54] In breast cancer patient, people who have low serum adiponectin levels tended to have more aggressive tumor.[73] The inverse relationship between serum adiponectin and endometrial cancer risk was also identified by two case control studies in Italy and Greece, respectively.[21,84] Adiponectin was found to be lower in prostate cancer patient comparing to healthy controls the levels were negatively correlated with histologic grade and disease stage.[37] Studies by Ishikawa et al.[49] showed that in gastric cancer patients, their plasma adiponectin levels were significantly lower than healthy controls. In addition, the plasma adiponectin was negatively associated with tumor size, depth of invasion and tumor stage in undifferentiated gastric cancer.[49] A prospective nested case-control study conducted by Wei et al.[122] observed that men with low plasma adiponectin levels had a higher risk of colon cancer than men with higher adiponectin. However, one study reported adiponectin was not associated with colorectal cancer.[67] Overall, studies in human subjects provided some evidence that adiponectin could protect against certain type of cancer.

The cancer preventive effect of adiponectin may partially be explained by its ability in modulating the biology of tumor cells. Studies by Yokota et al.,[131] found that adiponectin suppressed the growth of myelocyte cells, induced apoptosis in myelotye leukemia cells, and inhibited TNF-alpha production. Adiponectin was found to inhibit breast cancer MDA-MB-231 and MCF-7 cells proliferation and induce cell cycle arrest and apoptosis in these cells.[54,23] Bub et al.,[9] reported that adiponectin suppressed the growth of prostate cancer cells. In colon cancer cells, however, Ogunwobi et al.[79] demonstrated that adiponectin was a promoter of colon cancer HT29 cells.

Studies showed that plasma adiponectin level was negatively associated with obesity, glucose and lipid levels, and insulin resistance.[7] Weight control through dietary calorie restriction and/or exercise seems to elevate plasma adiponectin; however, the results are not very conclusive. Studies showed calorie restricted rats had a high level of plasma adiponectin with reduced blood glucose, plasma insulin, and triglyceride levels when compared with ad libitum-fed controls.[134] However, in a human study, the serum concentration of adiponection was not found to change in people after three weeks calorie restriction.[4] For the effect of exercise, Jamurtas et al.[51] showed that plasma adiponectin was not changed in people up to 48 hours post-acute exercise. Oberbach et al.[78] reported that after four weeks of physical training, adiponectin levels was significantly increased in people who had type 2 diabetes. The changing of adiponectin levels was correlated with enhanced insulin sensitivity.[78] Other adipocytokins, such as resistin and omentin, may also play a role in weight control-mediated cancer preventive effects, but their cellular and physiological function are still not clear.[71,129]

4. Other Hormones Related to Cancer Prevention by Weight Control

4.1. *Insulin*

Insulin is an important hormone that regulates blood glucose level. In the liver, it promotes glycogen synthesis by stimulating glycogen synthase and inhibiting glycogen phosphorylase. In muscle and fat tissues, insulin induces uptake of glucose via increasing GLUT4 expression. Insulin is also functioned as a moderate mitogen. After binding to its receptor, insulin may activate signaling pathways via phosphorylation of the insulin-receptor stubstarate-1, Akt, mitogen activated protein (MAP) kinase, and PI3K kinase.[92,27] Therefore, insulin has been found to induce the growth of both normal and cancerous cells.[121,61,6] Insulin also promotes the bioactivity of IGF-1 either via increasing the number of growth hormone receptors in the liver or reducing hepatic secretion of IGFBP1, which binds and inhibits the activity of IGF-1.[117,86]

Obesity or lack of physical activity is found to be a major factor inducing insulin resistance and further hyperinsulineamina. Epidemiological studies showed that increased plasma insulin was associated with a high risk of cancer.[53,77] Dietary calorie restriction and/or regular exercise has been linked with a decreased plasma insulin in several studies.[135,32,36] It is noted that weight control via decreasing calorie intake or increasing energy expenditure can regulate glucose homeostasis and increase insulin sensitivity.

4.2. *Glucocorticoids*

Glucocorticoid hormones are a class of steroid hormones. The major function of these hormones are involved in regulation of glucose metabolism, such as stimulation of gluconeogenesis in the liver, inhibition of glucose uptake in the muscle and adipose tissue, stimulation of fat breakdown in adipose tissue, and mobilization of amino acids from extrahepatic tissues. Glucocorticoids are also important in fetal development and have anti-inflammatory and immunosuppressive effects.

Glucocorticoid hormones act through binding to intercellular glucocorticosteroid receptor. After binding with the hormone, the new formed receptor-ligand complex dissociates with heat shock proteins and then translocates into the nucleus, where it binds again to glucocorticoid response elements (GRE) and acts as a transcription factor. Glucocorticoid receptor usually works as a negative transcription factor, and it has been shown to inhibit the transcription of almost all immune system-related genes. In some cases, activated glucocorticoid receptor may interfere with other transcription factors, such as AP-1 and NFκB[102,103] that are crucial in the regulation of a number of genes involved in inflammation, differentiation, cell proliferation, apoptosis, oncogenesis, and other biological processes.[58,96,35,57]

In addition, glucocorticoid steroids are potential tumor inhibitors. Administration of hydrocortisone in the diet showed preventive effect on the promoting phase of skin carcinogenesis in the mice.[112] There are a number of studies shown that glucocorticoid steroids are elevated in calorie restricted animals.[83,128] Adrenalectomy was found to decrease plasma corticosterone levels and abrogate the preventive effect of dietary restriction on skin tumor development.[83,101] Similar results were also observed in lung carcinogenesis but not in mammary gland tumors.[82,52] When administrating corticosterone in adrenalectomized mice, the cancer preventive effects of calorie restriction on skin carcinogenesis were restored as shown by our previous publication from the Birt lab.[101] Overall, the published data indicate glucocorticoids may be critical mediators in cancer prevention by calorie restriction.

5. Molecular Targets of Cancer Prevention by Weight Control

5.1. *Effects on cellular processes*

It is well known that cancer arises due to the loss of a normal growth control. In normal tissues, cell growth and cell death are highly regulated and balanced. In cancer, this regulation is disrupted, which is either from increased cell proliferation or loss of programmed cell death, or both.

The effect of calorie restriction on cell proliferation has been investigated in numerous studies. Lok *et al.*[66] reported that 25% of calorie restriction decreased cell proliferation by 72% in mammary gland and 30–60% in skin, esophagus, bladder, and GI tract of female Swiss Webster mice. Pashko and Schwartz[83] showed that 27% of food restriction suppressed TPA-induced epidermal [^{3}H]-thymidine incorporation. In C57BL/6×C3HF$_1$ mice, a murine strain that develops liver tumor spontaneously, 40% of dietary restriction was found to decrease cell proliferation significantly in the liver.[50] Dunn *et al.*[25] demonstrated that 20% of calorie restriction significantly inhibited BrdU incorporation in the bladders of p53 knock out mice. Restoration of IGF-1 brought the cell proliferation back to the level of the control mice.[25] Comparing to *ad libitum* feeding, 30% of dietary calorie restriction significantly inhibited cell proliferation in carcinogen treated mouse skin.[138] Using a heavy water labeling, Hsieh *et al.*[43] investigated a time-course of the effects of calorie restriction on cell proliferation rates in female C57BL/6J mice. It showed that the proliferation rates of mammary epithelial cells and T cells were markedly reduced within 2 weeks with calorie restriction regimen when compared to that of *ad libitum*-fed mice. Two weeks after refeeding, the cell proliferation rates rebounded to the basal level.[43] We found that the percentage of PCNA in skin epithelial cells was significantly lower in 20% of calorie restricted mice than *ad libitum*-fed mice, as shown by immunohistochemistry staining.[127] The percentage of the splenocyte in S phase was significantly reduced by 40% of calorie restriction in p53 knock out mice as well as wild type mice, as shown by Hursting *et al.*[139] Studies in Thompson lab showed that cell cycle regulators, i.e. cyclin D1, cyclin E, cyclin-dependent kinase (CDK)-2, and CDK-4, were decreased by 40% of calorie restriction in rat mammal carcinomas, while cyclin-dependent kinase inhibitors (CKI), i.e. Kip1/p27 and Cip1/p21, increased.[52,140] Overall, the effects of calorie restriction on cell proliferation are clear and reproducible in the animal models. For physical activity, studies in our lab found the cell proliferation rates of exercised mice with paired feeding had a lower rate than sedentary mice, but exercise with *ad libitum* feeding actually enhanced proliferative rates in epidermal cells, suggesting exercise alone without dietary calorie limitation might promote cellular proliferation and result in inconsistent impact on cancer protection.[127]

Programmed cell death or apoptosis is highly regulated by a series of arranged morphological and biochemical events.[1] It is important for maintenance of tissue homeostatsis, embryo development, and immune defense. Defects in apoptosis are thought to play an important role in cancer development.[34,142] Induction of apoptosis was observed in both normal liver and putative preneoplastic foci induced by hepatomitogen cyproerone acetate in dietary calorie restricted rats.[38] Dietary restriction was also found to induce apoptosis in the liver of C57BL/6×C3HF$_1$ mice.[50] Increased apoptosis was observed in the bladder preneoplasia of p-cresidine-treated p53-deficient mice by dietary calorie restriction.[25] In mammary gland, calorie restriction induced apoptosis in both premaglignant and malignant pathologies.[109] Thompson *et al.*, reported that apoptosis regulatory molecules, i.e. Bcl-2, Bcl-xl, and XIAP, decreased and Bax and Apaf-1 increased significantly in the mammary carcinomas of calorie restricted rats when compared with that of the control rats.[109] They also reported that the activities of both caspases-9 and caspases-3 were significantly induced and Akt phosphorylation was depressed by calorie restriction. The authors proposed that an induction of apoptosis by calorie restriction might be associated with its inhibitory effect on IGF-1 signaling. As for the physical activity, the research on apoptosis is sparse. Studies from our lab showed that caspase-3 activity but not caspase-3 protein increased significantly in epidermis of dietary calorie restricted and treadmill exercised mice in comparison with the sedentary controls.[127]

Collectively, all these data above indicate that modulation of cellular processes including inhibition of cell proliferation and restoration of apoptosis, is a molecular target in weight control for cancer prevention. Figure 1 shows a proposed mechanism by which weight control may inhibit cancer development via inhibiting the crosstalk between hormone-dependent and TPA-promoted signaling pathways, resulting in modulating cellular proliferation and anti-apoptosis. Weight loss reduces circulated growth factor and/or hormone levels such as IGF-1, leptin, and adiponectin, but enhances glucocorticoids, that thus inactivate TPA-induced signaling through hormone or growth factor-dependent cascades, e.g. Ras-MAPK and PI3K-Akt pathways. Finally, it may lead to an inhibition of TPA-induced cellular proliferation and elimination of IGF-1-persuaded anti-apoptosis.

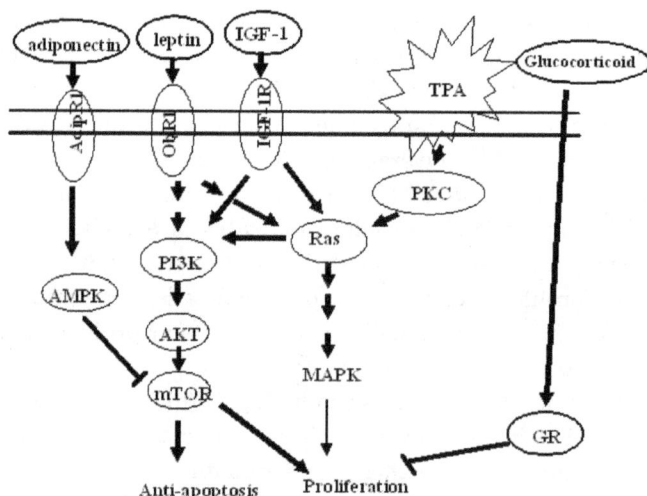

Fig. 1. A proposed mechanism by which weight control may inhibit cancer development via inhibiting the cross-talk between hormone-dependent and TPA-promoted signaling pathways, resulting in modulating cellular proliferation and anti-apoptosis.

5.2. *Reduction of oxidative stress*

Oxidative stress may injure cellular DNA, protein, and lipids in the tissue. It is thus associated with ageing and many chronic diseases. In carcinogenesis, reactive oxygen and nitrogen species can attack DNA directly and induce DNA mutations. Oxidative stress also occurs by the reactive products of a peroxidation from various macromolecules, such as lipid peroxidation that may lead to protein and DNA modification. Cumulative evidence has been shown that long-term calorie restriction in rodents extends maximum life span and decreases oxidative damage to DNA and proteins (reviewed by R. Gredilla *et al.*[141]). Qu *et al.*[87] found that 60% of calorie restriction completely abolished the increased oxidative damage in cloribrate-induced mouse liver.[87]

5.3. *Other possible impact*

Calorie restriction may also interfere with the expression balance between oncogene and tumor suppressor gene directly. "Oncogenes refer to genes whose activation can contribute to the development of cancer."[80] They are mutated versions from pro-oncogenes, which function in cell proliferation and differentiation. Overexpression of oncogenes usually causes out of control cellular growth. One of the most known oncogenes is Ras (Retrovirus-associated DNA sequences) family. Ras plays an important role in cell proliferation and can inactive tumor suppressors and promote cancer development.[26] "Tumor suppressor genes refer to those genes whose loss of function results in the promotion of malignancy."[80] Typically, a normal function of tumor suppressor genes is to inhibit cellular proliferation. Mutations of these genes usually result in a loss of their growth inhibition ability, which in turn may favor of cellular proliferation. Some examples of well-known tumor suppressor genes include p53, retinoblastoma susceptibility gene, Wilms' tumors, neurofibromatosis type-1, and familial adenomatosis polyposis coli, etc. Previous studies have demonstrated that food restriction may induce an over-expression of tumor suppressor gene p53.[26] In the Brown-Norway rats fed with calorie restricted diet or *ad libitum* diet, Hass *et al.*[41] found that pancreatic acinar cells from calorie restricted animals had a lower growth rate and less N-methyl-N'-nitro-N-nitrosoguanidine (MNNG)-induced transformation. Calorie restriction derived cells showed decreased c-Ha-ras gene expression, lower rate of mutation of p53 tumor suppressor gene, and increased genomic methylations of DNA.[41]

DNA repair occurs in the normal mammalian cells to repair DNA damage caused by multiple factors including oxidative stress. It reported that dietary calorie restriction enhanced DNA repair ability against DNA damage caused by UV exposure.[64,123] Hursting *et al.*[45] suggested that both calorie restriction and exercise could induce DNA repair pathway, therefore block the early stage of carcinogenesis. However, 30% of calorie restriction failed to activate DNA repair pathways and inhibit tumor development in the DNA mismatch deficient mice.[114]

6. Summary

As we know more about the protective mechanisms of weight control, it is becoming apparent that it is not only one single mechanism involved. Most likely, it is a combination of multiple factors and multiple signaling pathways involved. Hundreds of biological molecules may cooperate in this network complex. Therefore, traditional molecular biology techniques seem not to meet the requirement to gain a broader and deeper overview of the mechanisms. Fortunately, recently developed technologies named "-omics" may provide us a chance to take a global view of these biological processes. The "-omics" techniques such as genomics, proteomics, and lipidomics, etc., usually present a profile change of gene, protein, and lipid expression, respectively. Microarray study is the first step to obtain a gene expression profile, together with proteomics and lipidomics may generate a clear picture of a profiling response to the weight loss treatment. Our on-going studies by using these state of the art technologies hopefully lead to a better and deeper mechanistic information regarding the established cancer prevention by weight control in the near future.

Acknowledgments

This work was supported by an Innovative Research Grant from the Terry C. Johnson Center for Basic Cancer Research, Kansas State University, NIH COBRE Award P20 RR15563 and Kansas State matching support, and NIH R01 CA106397. This is a journal contribution #09-017-J by the Kansas Agricultural Experiment Station, Kansas State University.

References

1. J. M. Adams, *Genes Dev.* **17**, 2481 (2003).
2. Y. Arita *et al.*, *Circulation* **105**, 2893 (2002).
3. W. A Banks, A. J. Kastin, W. Huang, J. B. Jaspan and L. M. Maness, *Peptides* **17**, 305 (1996).
4. K. E. Barnholt *et al.*, *Am. J. Physiol. Endocrinol. Metab.* **290**, E1078 (2006).
5. L. Basterfield, L. J. M. Reul and J. C. Mathers, *J. Nutr.* **135**, 3002S (2005).
6. J. Björk, J. Nilsson, R. Hultcrantz and C. Johansson, *Scand. J. Gastroenterol.* **28**, 879 (1993).
7. M. Blüher *et al.*, *J. Clin. Endocrinol. Metab.* **91**, 2310 (2006).
8. D. K. Bol, K. Kiguchi, I. Gimenez-Conti, T. Rupp and J. DiGiovanni, *Oncogene* **14**, 1725 (1997).
9. J. D. Bub, T. Miyazaki and Y. Iwamoto, *Biochem. Biophys. Res. Commun.* **340**, 1158 (2006).
10. L. A. Campfield, F. J. Smith, Y. Guisez, R. Devos and P. Burn, *Science* **269**, 546 (1995).
11. F. Canzian *et al.*, *Br. J. Cancer* **94**, 299 (2006).
12. J. F. Caro *et al.*, *Lancet* **348**, 159 (1996).
13. A. Cats *et al.*, *Cancer Res.* **56**, 523 (1996).
14. CDC, *MMWR Morb Mortal Wkly Rep.* **55**, 985 (2006).
15. J. M. Chan *et al.*, *Science* **279**, 563 (1998).
16. S. Chang *et al.*, *Prostate* **46**, 62 (2001).
17. D. C. Chen *et al.*, *Cancer Lett.* **237**, 109 (2006).
18. A. P. Chokkalingam *et al.*, *Prostate* **52**, 98 (2002).
19. D. R. Clemmons, *Mol. Reprod. Dev.* **35**, 368 (1993).
20. J. Cornish *et al.*, *J. Endocrinol.* **175**, 405 (2002).
21. M. L. Dal *et al.*, *J. Clin. Endocrinol. Metab.* **89**, 1160 (2004).
22. T. M. DeChiara, A. Efstratiadis and E. J. Robertson, *Nature* **345**, 78 (1990).
23. M. N. Dieudonne *et al.*, *Biochem. Biophys. Res. Commun.* **345**, 271 (2006).
24. J. DiGiovanni *et al.*, *Cancer Res.* **60**, 1561 (2000).
25. S. E. Dunn *et al.*, *Cancer Res.* **57**, 4667 (1997).
26. G. Fernandes *et al.*, *Proc. Natl. Acad. Sci. USA* **92**, 6494 (1995).
27. C. A. Finlayson *et al.*, *Metabolism* **52**, 1606 (2003).
28. F. F. Fisher *et al.*, *Diabetologia* **48**, 1084 (2005).
29. L. Fontana, S. Klein and J. O. Holloszy, *Am. J. Clin. Nutr.* **84**, 1456 (2006).
30. K. A. Frankenberry *et al.*, *Int. J. Oncol.* **28**, 985 (2006).
31. R. C. Frederich *et al.*, *Nat. Med.* **1**, 1311 (1995).
32. C. M. Friedenreich and M. R. Orenstein, *J. Nutr.* **132**, 3456S (2002).
33. C. Garofalo and E. Surmacz, *J. Cell Physiol.* **207**, 12 (2006).
34. R. Gerl and D. L. Vaux, *Carcinogenesis.* **26**, 263 (2005).
35. S. Ghosh and M. Karin, *Cell* **109**, S81 (2002).
36. E. Giovannucci, *J. Nutr.* **131**, 3109S (2001).
37. S. Goktas *et al.*, *Urology* **65**, 1168 (2005).
38. B. Grasl-Kraupp *et al.*, *Proc. Natl. Acad. Sci. USA* **91**, 9995 (1994).
39. J. L. Halaas *et al.*, *Science* **269**, 543 (1995).
40. J. C. Hardwick *et al.*, *Gastroenterology* **121**, 79 (2001).
41. B. S. Hass, R. W. Hart, M. H. Lu and B. D. Lyn-Cook, *Mutat. Res.* **295**, 281 (1993).
42. J. M. Holly *et al.*, *Growth Regul.* **3**, 88 (1993).
43. E. A. Hsieh, C. M. Chai and M. K. Hellerstein, *Am. J. Physiol. Endocrinol. Metab.* **288**, E965 (2005).

44. L Huang and C. Li, *Cell Res.* **10**, 81 (2000).
45. S. D. Hursting and F. W. Kari, *Mutat. Res.* **443**, 235 (1999).
46. S. D. Hursting, J. A. Lavigne, D. Berrigan and S. N. Perkins, *Annu. Rev. Med.* **54**, 131 (2003).
47. S. D. Hursting *et al.*, *Cancer Res.* **57**, 2843 (1997).
48. S. D. Hursting, B. R. Switzer, J. E. French and F. W. Kari, *Cancer Res.* **53**, 2750 (1993).
49. M. Ishikawa *et al.*, *Clin. Cancer Res.* **11**, 466 (2005).
50. S. J. James and L. Muskhelishvili, *Cancer Res.* **54**, 5508 (1994).
51. A. Z. Jamurtas *et al.*, *Eur. J. Appl. Physiol.* **97**, 122 (2006).
52. W. Jiang, Z. Zhu, J. N. McGinley and H. J. Thompson, *J. Nutr.* **134**, 1152 (2004).
53. R. Kaaks, *Novartis Found. Symp.* **262**, 247 (2004).
54. J. H. Kang *et al.*, *Arch. Pharm. Res.* **28**, 1263 (2005).
55. K. Kantartzis *et al.*, *Obes. Res.* **13**, 1683 (2005).
56. F. W. Kari, S. E. Dunn, J. E. French and J. C. Barrett, *J. Nutr. Health Aging* **3**, 92 (1999).
57. M. Karin *et al.*, *Nat. Rev. Cancer.* **2**, 301 (2002).
58. M. Karin and L. Chang, *J. Endocrinol.* **169**, 447 (2001).
59. J. Khosravi, A. Diamandi, J. Mistry and A. Scorilas, *J. Clin. Endocrinol. Metab.* **86**, 694 (2001).
60. O. Y. Kim *et al.*, *Clin. Chim. Acta.* **370**, 63 (2006).
61. M. Koenuma, T. Yamori and T. Tsuruo, *Jpn. J. Cancer Res.* **80**, 51 (1989).
62. R. T. Kurmasheva and P. J. Houghton, *Biochim. Biophys. Acta.* **1766**, 1 (2006).
63. P. Lagiou *et al.*, *Int. J. Cancer* **76**, 25 (1998).
64. J. M. Lipman, A. Turturro and R. W. Hart, *Mech. Ageing Dev.* **48**, 135 (1989).
65. Z. Liu, T. Uesaka, H. Watanabe and N. Kato, *Int. J. Oncol.* **19**, 1009 (2001).
66. E. Lok *et al.*, *Cancer Lett.* **51**, 67 (1990).
67. A. Lukanova *et al.*, *Cancer Epidemiol. Biomarkers Prev.* **15**, 401 (2006).
68. M. Maffei *et al.*, *Nat. Med.* **1**, 1155 (1995).
69. V. Mai, L. H. Colbert, D. Berrigan and S. N. Perkins, *Cancer Res.* **63**, 1752 (2003).
70. S. Margetic, C. Gazzola, G. G. Pegg and R. A. Hill, *Int. J. Obes. Relat. Metab. Disord.* **26**, 1407 (2002).
71. P. G. McTernan, C. M. Kusminski and S. Kumar, *Curr. Opin. Lipidol.* **17**, 170 (2006).
72. T. Miyazaki *et al.*, *Biochem. Biophys. Res. Commun.* **333**, 79 (2005).
73. Y. Miyoshi *et al.*, *Clin. Cancer Res.* **9**, 5699 (2003).
74. T. Moore *et al.*, *Cancer Res.* **68**, 3680 (2008).
75. C. Z. Moreschi, *Immunitaetsforsch.* **2**, 651 (1909).
76. D. M. Muoio and G. Lynis Dohm, *Best Pract. Res. Clin. Endocrinol. Metab.* **16**, 653 (2002).
77. T. I. Nilsen and L. J. Vatten, *Br. J. Cancer* **84**, 417 (2001).
78. A. Oberbach *et al.*, *Eur. J. Endocrinol.* **154**, 577 (2006).
79. O. O. Ogunwobi and I. L. Beales, *Regul. Pept.* **134**, 105 (2006).
80. C. Osborne, P. Wilson and D. Tripathy, *Oncologist* **9**, 361 (2004).
81. G. Pandini *et al.*, *Int. J. Biol. Chem.* **277**, 39684 (2002).
82. L. L. Pashko and A. G. Schwartz, *Carcinogenesis* **17**, 209 (1996).
83. L. L. Pashko and A. G. Schwartz, *Carcinogenesis* **13**, 1925 (1992).
84. E. Petridou *et al.*, *J. Clin. Endocrinol. Metab.* **88**, 993 (2003).
85. Z. Pietrzkowski *et al.*, *Mol. Cell Biol.* **12**, 3883 (1992).
86. D. R. Powell *et al.*, *J. Biol. Chem.* **266**, 18868 (1991).

87. B. Qu *et al.*, *FEBS Lett.* **473**, 85 (2000).
88. B. S. Reddy, S. Sugie and A. Lowenfels, *Cancer Res.* **48**, 7079 (1988).
89. A. G. Renehan *et al.*, *Lancet.* **63**, 1346 (2004).
90. O. Rho *et al.*, *Mol. Carcinog.* **17**, 62 (1996).
91. A. A. Richards *et al.*, *Mol. Endocrinol.* **20**, 1673 (2006).
92. D. W. Rose *et al.*, *Oncogene* **17**, 889 (1998).
93. B. A. Ruggeri, D. M. Klurfeld, D. Kritchevsky and R. W. Furlanetto, *Cancer Res.* **49**, 4130 (1989).
94. K. Saglam, E. Aydur, M. Yilmaz and S. Göktaş, *J. Urol.* **169**, 1308 (2003).
95. C. Sell, M. Rubini, R. Rubin, J. P. Liu and A. Efstratiadis, *Proc. Natl. Acad. Sci. USA* **90**, 11217 (1993).
96. E. Shaulian and M. Karin, *Oncogene* **20**, 2390 (2001).
97. G. E. Sonnenberg, G. R. Krakower and A. H. Kissebah, *Obes. Res.* **12**, 180 (2004).
98. M. A. Soos, C. E. Field and K. Siddle, *Biochem. J.* **290**, 419 (1993).
99. P. Stattin *et al.*, *Cancer Epidemiol. Biomarkers Prev.* **12**, 474 (2003).
100. P. Stattin *et al.*, *J. Clin. Endocrinol. Metab.* **86**, 1341 (2001).
101. J. W. Stewart *et al.*, *Carcinogenesis* **26**, 1077 (2005).
102. E. Stöcklin, M. Wissler, F. Gouilleux and B. Groner, *Nature* **383**, 726 (1996).
103. N. Subramaniam, J. Campión, I. Rafter and S. Okret, *Biochem. J.* **370**, 1087 (2003).
104. M. Sulkowska, J. Golaszewska, A. Wincewicz, M. Koda, M. Baltaziak and S. Sulkowski, *Pathol. Oncol. Res.* **12**, 69 (2006).
105. M. Takahashi *et al.*, *Int. J. Obes. Relat. Metab. Disord.* **24**, 861 (2000).
106. A. Tannenbaum, *Ann. NY Acad. Sci.* **49**, 5 (1947).
107. S. Teramukai *et al.*, *Jpn. J. Cancer Res.* **93**, 1187 (2002).
108. H. J. Thompson, Z. Zhu and W. Jiang, *Cancer Res.* **64**, 1541 (2004).
109. H. J. Thompson, Z. Zhu and W. Jiang, *J. Nutr.* **134**, 3407S (2004).
110. E. B. Thorling, N. O. Jacobsen and K. Overvad, *Eur. J. Cancer Prev.* **2**, 77 (1993).
111. S. P. Thumiger *et al.*, *Growth Factors* **23**, 151 (2005).
112. N. Trainin, *Cancer Res.* **23**, 415 (1963).
113. J. V. Tricoli *et al.*, *Cancer Res.* **46**, 6169 (1986).
114. J. L. Tsao *et al.*, *Carcinogenesis* **23**, 1807 (2002).
115. H. Tsuchihashi *et al.*, *J. Surg. Res.* **134**, 348 (2006).
116. A. Ullrich *et al.*, *EMBO J.* **5**, 2503 (1986).
117. L. E. Underwood *et al.*, *Horm. Res.* **42**, 145 (1994).
118. M. Van Heek *et al.*, *Horm. Metab. Res.* **28**, 653 (1996).
119. M. von Eynatten *et al.*, *Clin. Chem.* **52**, 853 (2006).
120. D. W. Voskuil *et al.*, *Cancer Epidemiol. Biomarkers Prev.* **14**, 195 (2005).
121. L. F. Watkins, L. R. Lewis and A. E. Levine, *Int. J. Cancer* **45**, 372 (1990).
122. E. K. Wei *et al.*, *J. Natl. Cancer Inst.* **97**, 1688 (2005).
123. N. Weraarchakul, R. Strong, W. G. Wood and A. Richardson, *Exp. Cell Res.* **181**, 197 (1989).
124. E. Wilker *et al.*, *Mol. Carcinog.* **25**, 122 (1999).
125. E. Wilker *et al.*, *Mol. Carcinog.* **44**, 137 (2005).
126. Y. Wu *et al.*, *Cancer Res.* **63**, 4384 (2003).
127. L. Xie *et al.*, *J. Biol. Chem.* **282**, 28025 (2007).
128. A. L. Yaktine, R. Vaughn, D. Blackwood, E. Duysen and D. F. Birt, *Mol. Carcinog.* **21**, 62 (1998).
129. R. Z. Yang *et al.*, *Am. J. Physiol. Endocrinol. Metab.* **290**, E1253 (2006).
130. D. Yee *et al.*, *Mol. Endocrinol.* **3**, 509 (1989).
131. T. Yokota *et al.*, *Blood* **96**, 1723 (2000).

132. Y. Zhang *et al.*, *Nature* **372**, 425 (1994).
133. H. Zhao *et al.*, *J. Urol.* **169**, 714 (2003).
134. M. Zhu *et al.*, *Exp. Gerontol.* **39**, 1049 (2004).
135. Z. Zhu, W. Jiang, J. McGinley, P. Wolfe and H. J. Thompson, *Mol. Carcinog.* **42**(3), 170 (2005).
136. T. K. Raushan and P. J. Houghton, *Biochem. Biophys. Acta* **1766**, 1 (2006).
137. A. Koerner, J. Kratzsch and W. Kiess, *Best Pract. Res. Clin. Endocrinol. Metab.* **19**, 525 (2005).
138. W. H. Fischer and W. K. Lutz, *Toxicol. Lett.* **98**, 59 (1998).
139. S. D. Hursting, S. N. Perkins and J. M. Phang, *Proc. Natl. Acad. Sci. USA* **91**, 7036 (1994).
140. Z. Zhu, W. Jiang and H. J. Thompson, *Carcinogenesis* **24**, 1225 (2003).
141. R. Gredilla and G. Barja, *Endocrinology* **146**, 3713 (2005).
142. L. Lossi, C. Cantile, I. Tamagno and A. Merighi, *Vet. J.* **170**, 52 (2005).

Keyword Index

adhesion, 125–127, 129, 131
androgen deprivation therapy, 144

biomechanics, 38
breakdown of coherent states, 172, 174

cancer diagnosis, 88, 91
cancer electrodynamics, 172, 173, 176
cancer prevention, 191, 192, 195–197, 200, 201, 203
cell-cycle pathway, 3
chemotherapy, 1–3, 10, 11, 13, 14, 16
collagen, 79–81, 84
contractile forces, 25–28, 32, 34
cytoskeletal remodeling, 26, 27, 37

dietary calorie restriction, 191, 195, 198, 200, 201, 203
differential scatter cross-section, 44
Droop equation, 143

fiber diffraction, 79, 84
focal adhesions, 26, 28, 33, 34
fractional kinetics, 55, 57, 62

glioma, 55–57, 63, 64, 67, 69
glucocorticoids, 191, 200, 201

hERG1, 109, 112–116, 121
humans, 161
hypoxia, 2, 3, 7

IGF-1, 191–196, 199–201
incidence, 159–167
integrin, 125–130, 133
integrins, 23–25, 27–29, 33, 36, 111
ion channels, 109–113, 116, 121

keratin, 89

leptin, 191, 196, 197, 201

mathematical modelling, 115, 122
mechano-stimuli, 125, 127
migration–proliferation dichotomy, 56, 57, 60, 63, 65, 66, 68, 69
model, 139, 141–145, 147, 148, 150–152, 154, 155
Monte Carlo radiation transport, 44, 47, 48, 53
multi-scale cancer modeling, 97, 98
multiscale modelling, 1–3
mutation, 159, 165, 167

neoplasms, 159

p53, 109, 110, 114, 115, 117–122
physical activity, 191, 192, 195, 199, 201
prostate cancer, 139, 140, 142, 152, 155
prostate specific antigen, 139, 140

radiation biophysics, 43, 53
radiation therapy, 43, 44, 49, 52, 53
reactive oxygen species, 125, 128
RNAi, 27

scaling law, 104–106
signaling, 125–133
stiffness, 25–27, 29, 30, 36–38

tumor growth, 101, 103, 106

viscoelasticity, 30

Warburg effect, 172, 174, 179, 180, 184–186
weight control, 191–193, 195, 197, 199, 201–203

X-rays, 78, 79

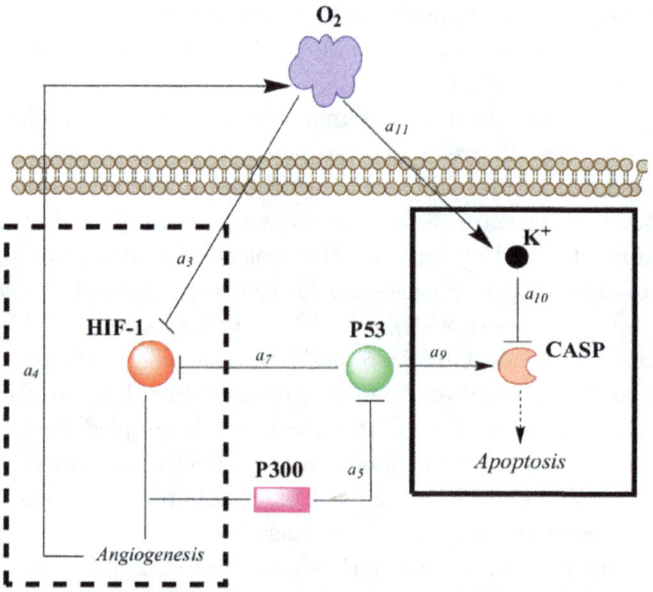

$$\text{(i)} \quad \varnothing \xrightarrow{a_{hif}} \text{HIF-1}$$

$$\text{(ii)} \quad \varnothing \xrightarrow{a_{o2}} O_2$$

$$\text{(iii)} \quad \varnothing \xrightarrow{a_8} \text{p300}$$

$$\text{(iv)} \quad \varnothing \xrightarrow{a_{p53}} \text{P53}$$

$$\text{(v)} \quad \varnothing \xrightarrow{a_{12}} \text{CASP}$$

$$\text{(vi)} \quad \text{CASP} \xrightarrow{a_{13}} \varnothing$$

$$\text{(vii)} \quad \text{K} \xrightarrow{a_{14}} \varnothing$$

mitochondria   NOX   other signals

www.ingramcontent.com/pod-product-compliance
Lightning Source LLC
Chambersburg PA
CBHW081501190326
41458CB00015B/5301